Christoph Kunert

Design for Stability in Transport Logistics

Definition, Concepts and Evaluation

WISSENSCHAFTLICHE BERICHTE

Institut für Fördertechnik und Logistiksysteme
am Karlsruher Institut für Technologie (KIT)

BAND 91

Design for Stability in Transport Logistics

Definition, Concepts and Evaluation

by
Christoph Kunert

KIT Scientific Publishing

Dissertation, Karlsruher Institut für Technologie
KIT-Fakultät für Maschinenbau

Tag der mündlichen Prüfung: 25. Mai 2018
Gutachter: Prof. Dr.-Ing. Kai Furmans, Univ.-Prof. Dr. Michael Henke

Impressum

KIT Scientific Publishing

Karlsruher Institut für Technologie (KIT)
KIT Scientific Publishing
Straße am Forum 2
D-76131 Karlsruhe

KIT Scientific Publishing is a registered trademark
of Karlsruhe Institute of Technology.
Reprint using the book cover is not allowed.

www.ksp.kit.edu

Print on Demand 2018 – Gedruckt auf FSC-zertifiziertem Papier

ISSN 0171-2772
ISBN 978-3-7315-0806-9
DOI 10.5445/KSP/1000083491

Design for Stability in Transport Logistics

Definition, Concepts and Evaluation

zur Erlangung des akademischen Grades eines

Doktors der Ingenieurswissenschaften

von der Fakultät für Maschinenbau
des Karlsruher Instituts für Technologie (KIT)

genehmigte

Dissertation

von

Christoph Kunert

aus Edertal

Tag der mündlichen Prüfung: 25. Mai 2018
Erster Gutachter: Prof. Dr.-Ing. Kai Furmans
Zweiter Gutachter: Univ.-Prof. Dr. Michael Henke

Danksagung

Die vorliegende Arbeit entstand während meiner Tätigkeit als wissenschaftlicher Mitarbeiter am Institut für Fördertechnik und Logistiksysteme des Karlsruher Instituts für Technologie. Ich möchte mich an dieser Stelle bei allen Personen bedanken, die zum Gelingen der Arbeit beigetragen haben.

Meinem Doktorvater Prof. Dr.-Ing. Kai Furmans bin ich sehr dankbar, da er durch die Übernahme der Betreuung die Erstellung dieser Arbeit ermöglicht hat. Dabei genoss ich sehr viel Freiraum und auch viel Unterstützung. Univ.-Prof. Dr. Michael Henke danke ich für die Übernahme des Korreferats. Prof. Dr.-Ing. Michael Hoffmann danke ich für die Übernahme des Prüfungsvorsitzes.

Den aktiven und ehemaligen Kollegen am IFL danke ich für die tolle Arbeitsatmosphäre und den starken Zusammenhalt. Ob auf dem Skiseminar in Obersaxen, abends beim Grillen am IFL oder gemeinsam am Tischkicker – der Spaß kam niemals zu kurz. Dies hat nicht nur zum Gelingen der Arbeit beigetragen sondern machte den Lebensabschnitt am IFL zu einer unvergesslichen Zeit, an die ich mich sehr gerne zurückerinnern werde.

Besonderer Dank gilt meinen Freunden und meiner Familie, die auch in den arbeitsintensiven Phasen stets Verständnis für mich zeigten und mir damit großen Rückhalt boten. Meinen Eltern und meinem Bruder bin ich dankbar für die grenzenlose Unterstützung, die ich auf meinem bisherigen Lebensweg erfahren habe.

Mein ganz persönlicher Dank gilt meiner Partnerin Nicole. Sie hat mich stets ertragen und trotzdem unterstützt.

Karlsruhe, Juli 2018 Christoph Kunert

Kurzfassung

Lean Management steht für eine Sammlung an Methoden gepaart mit einer passenden Management-Philosophie mit dem Ziel, Verschwendung in Logistikprozessen zu eliminieren und die Produktivität zu erhöhen. Im Bereich der Produktionslogistik ist Lean bereits etabliert und wird in der Industrie erfolgreich umgesetzt. Forschungsergebnisse zeigen, dass Lean auch in anderen Teildisziplinen der Logistik, z.B. der Lagerlogistik, funktioniert. Dort erfreut sich Lean wachsender Beliebtheit und ist in der Industrie zunehmend weit verbreitet. Im Bereich der Transportlogistik ist der Reifegrad von Lean sowohl in Bezug auf den Stand der Forschung als auch in Bezug auf die Verbreitung in der Industrie gering. Die vorliegende Arbeit leistet einen Beitrag, um diese Lücke zu schließen.

Ein zentrales Element schlanker Logistiksysteme sind Designmaßnahmen, die zu einer Stabilisierung der Prozesse führen. Zum heutigen Zeitpunkt liegt jedoch keine umfassende, allgemein anwendbare Definition des Begriffs vor, obwohl er in der Literatur häufig verwendet wird. Daher wird im ersten Schritt der Arbeit eine Definition des Begriffs Stabilität im Kontext von Logistiksystemen entwickelt. Gemäß dieser Definition ist Stabilität die Wahrscheinlichkeit, dass ein Prozessergebnis innerhalb eines gewünschten Zielzustands liegt. Die Dimensionen, in denen Stabilität gemessen werden kann, sind Bestand, Kapazität und Zeit. Stabilisierung kann durch direktes Verringern der prozessinhärenten Variabilität oder durch Pufferung in einer Kombination von Bestand, Kapazität oder Zeit erzielt werden.

Als Maßnahme für ein auf Stabilität ausgerichtetes Systemdesign wird das Prinzip der Heijunka Nivellierung aus der Produktionslogistik auf die Transportlogistik übertragen. Kernidee des Konzepts ist es, Variabilität durch ein Wechselspiel aus Bestands- und Auftragspuffer aus der Kapazitätsdimension in die Bestandsdimension zu verschieben. Es wird ein Bestandsmodell vorgestellt, dass das Systemverhalten beschreibt und die Möglichkeit eröffnet, das für eine gewünschte statistische Sicherheit erforderliche Bestandsniveau zu berechnen. Weiterhin wird ein Optimierungsmodell vorgestellt, dass ein Nivellierungsmuster mit glatten Bestellmengen unter Beachtung von Kapazitätsrestriktionen berechnet. Beide Modelle werden in einen integrierten Planungsprozess eingebettet aus dem ein Plan for Every Part (PFEP) resultiert.

Um die Effektivität für unterschiedliche Mengen- und Gewichtsstrukturen zu untersuchen, wird ein agentenbasiertes Simulationsmodell eines Milkruns mit mehreren Rohteilen, die per nivellierender Bestellpolitik beschafft werden, entwickelt. Die Simulationsexperimente zeigen, dass Heijunka den gewünschten Effekt der Stabilisierung der erforderlichen Transportkapazität erzielt: Für eine gegebene statische Sicherheit kann mit geringerer Kapazität der gleiche Durchsatz wie bei einem gewöhnlichen Kanbansystem erzielt werden, da der erforderliche Puffer von der Kapazitätsdimension in die Bestandsdimension verschoben wird. Es wird gezeigt, dass ein längeres Nivellierungsmuster auf Transportebene zu einem stärkeren Nivellierungseffekt führt, da die reservierte Kapazität genauer der tatsächlichen Nachfrage entspricht. Auf Teileebene können längere Nivellierungsmuster zu schwankenden Bestellmengen führen und daher die Effektivität der Nivellierung verringern.

Der Effekt der Stabilisierung der erforderlichen Transportkapazität wird auf Kosten eines höheren Pufferbestandes erzielt. Um die Effizienz des Konzepts zu bewerten, modellieren wir die Betriebskosten, d.h. die Summe aus Bestands- und Kapazitätskosten, eines Heijunka-nivellierten Logistiksystems in Abhängigkeit von der Pufferallokation. Die Pareto-effiziente Pufferallokation ist durch das Minimum der Betriebskostenfunktion bzw. der Nullstellen deren Ableitung, der Grenzkostenfunktion, gegeben. Die kostenminimale Pufferallokation wird durch eine Enumeration über verschiedene Faktorkostenverhältnisse und durch eine Linearisierung der Grenzkostenfunktion durch eine Taylorreihe bestimmt. Das Modell zeigt, dass die Pareto-effiziente Pufferallokation für einen großen Bereich praktisch relevanter Faktorkostenverhältnisse nur durch Heijunka-Nivellierung erreicht werden kann.

Als numerisches Beispiel bewerten wir den Kompromiss zwischen Kapazitäts- und Bestandskosten am Beispiel des deutschen Automobilzulieferers ZF

Friedrichshafen. Dazu wird das Transportsystem des Beschaffungsnetzwerks mit den vorgestellten Methoden ausgelegt. Es werden fünf Milkrun-Touren gebildet, in deren Rahmen Material bei nahegelegenen Lieferanten abgeholt wird. Für jede Tour wird eine kostenminimale Pufferallokation bestimmt. Weiterhin werden die Transportkosten des Milkruns mit geglätteten Bestellmengen mit den Kosten des Gebietsspediteurs mit MRP-Bestellungen verglichen. Es wird gezeigt, dass die Transportkosten im Falle der nivellierten Bestellpolitik in Kombination mit Milkruns am Beispiel der Fallstudie geringer sind. Demzufolge führt eine Verschiebung der Variabilität aus der Kapazitätsdimension in die Bestandsdimension am Beispiel der Fallstudie zu einer Verringerung der Betriebskosten.

Abstract

Lean management describes a set of methods combined with a management philosophy which aims at eliminating waste in logistics processes to foster productivity. In the field of production logistics, lean is already widespread among industry practitioners and applied successfully. Research has shown that lean also works in the warehousing environment which led to an increasing popularity and increasing dissemination among industry practitioners. In transport logistics, lean is still at a low level of maturity in both research and practice. This thesis makes a contribution at closing this gap.

One central element of lean logistics systems are design measures, which lead to a stabilization of processes. Up to date, no uniform generally applicable definition of stability for logistics systems exists and is thus derived in this thesis. According to this definition, stability is the probability of a process outcome being within a desired target state. The dimensions in which we can measure stability are inventory, capacity and time. Stabilization can be achieved by directly decreasing process-inherent variability or by buffering by some combination of inventory, capacity or time.

As a measure of "Design for Stability", the principle of heijunka leveling is transferred from production logistics to transport logistics. The idea of this concept is to employ a combination of an inventory and an order buffer to move variability from the costly capacity dimension to the less costly inventory dimension. We propose an inventory model which describes the system's behavior and is suitable to calculate the required buffer inventory for a desired statistical safety. Moreover, we present an optimization model which calculates a leveled replenishment pattern under the restriction of a limited capacity. Both models are part of an integrated planning process which results in a Plan for Every Part (PFEP).

We create an agent based simulation model of a milk run system with leveled replenishment to evaluate the effectiveness in case of two different scenarios of weight and quantity structures. By conducting simulation experiments, we show that heijunka is an effective measure for stabilizing the required transport capacity: with less transport capacity we can achieve the same throughput as a regular kanban system because we require less capacity to buffer the varying consumption. We find that on transport level, a longer heijunka pattern results in a higher effectiveness of leveling. The reason is that the reserved capacity can be more easily adjusted to the demand requirement. On part level, a longer heijunka pattern can lead to varying order quantities and therefore diminished effectiveness.

The effect of stabilizing the required capacity is achieved at the expense of a higher inventory buffer. To evaluate the efficiency of the proposed "Design for Stability", we model the system operating costs, i.e. the sum of inventory and capacity costs, of a heijunka-levelled transport system as a function of the buffer allocation. The Pareto-efficient buffer allocation is given by the minimum of the total cost function or the zeroes of its derivative, the marginal cost function. We determine the location of the optimum by both an enumeration of different factor cost ratios and a linearization of the marginal cost function by means of a Taylor series. The model shows that for a wide range of practically relevant parameter combinations, the Pareto-efficient buffer allocation can only be achieved by heijunka leveling.

As a numerical example we investigate the trade-off between catch-up capacity and buffer inventory by the case study of the German automotive supplier ZF Friedrichshafen. We design the system and create five sample milk runs which procure the parts from adjacent suppliers. For each milk run, we calculate a cost-minimal buffer allocation in the sense of Pareto. Moreover, we compare the transport costs of the milk run in association with heijunka leveled replenishment to an area freight forwarder concept with MRP replenishment. We find that transport costs can be reduced by a leveled replenishment policy. In the case study, employing a leveled replenishment policy leads to lower total costs of operation for the system.

Table of Contents

List of Figures

List of Tables

List of Symbols

Abbreviations

AFF	Area Freight Forwarder
BOM	Bill of Materials
CEP	Courier and Express parcel service provider
cf	confer
CM	Continuous Model
CM-T	Continuous Model – Taylor Approximation
CT	Cycle time
CUC	Catch-Up Capacity
CV	Coefficient of Variation
DM	Discrete Model
EPEI	Every Part Every Interval
ERP	Enterprise Resource Planning
FTL	Full Truck Load
HEI	Heijunka
i.e.	id est
KA	Kanban
LSP	Logistics Service Provider
LTL	Less than Truck Load
MOQ	Minimum Order Quantity
MR	Milk run
MRP	Material Requirements Planning
OEE	Overall Equipment Effectiveness
PFEP	Plan For Every Part
SD	Standard Deviation
TPS	Toyota Production System
WIP	Work in process

Mathematical Symbols

b_i	Total quantity of part i in leveling period
B_t	Order backlog in period t
$\vec{c}$	Probability distribution of capacity
C_{CUC}	Catch-Up Capacity (absolute)
C_t	Capacity in period t
C_{Tour}	Capacity which is required for one tour
C_v	Capacity of vehicle
$C_{v,min}$	Capacity of vehicle with smallest possible payload
$C_{v,max}$	Capacity of vehicle with highest possible payload
CUC	Catch-up capacity (relative)
$\vec{d}$	Probability distribution of demand
D_t	Demand in period t
$\Delta_{rel}(CUC_{CM})$	Allocation inefficiency caused by the inaccuracy of the approximation of the continuous model (CM)
Δ_{cv^2}	Difference between cv^2 of orders and demand.
d_{Tour}	Total distance covered by tour
$E(X)$	Expected value of random variable X
f	Variability parameter
i	Inventory holding cost rate in the leveling horizon (interest rate)
I_{phys}	Physical inventory
I_t	Physical inventory in period t
$k_{duration}$	Unit cost rate per duration
$k_{distance}$	Unit cost rate per distance
k_f	Fixed costs of transportation of one tour (Strategy A)
k_f'	Fixed costs of transportation of one tour (Strategy B)
$k_{h,i}$	Inventory holding costs of one unit of inventory of part i in the leveling horizon
k_v	Variable costs of transportation of one tour
K_I	Inventory Costs in the leveling horizon
K_{TO}	Total Costs

K_{TR}	Transport costs in the leveling horizon
L	Probability distribution of lead time
N_D	Number of days in leveling horizon (=days on which there is a demand)
N_K	Number of kanbans
N_P	Number of different parts that are served by the tour
N_{PC}	Number of parameter combinations
N_T	Number of tours in leveling horizon (=days on which there is capacity)
$N_{T,max}$	Maximum number of tours in leveling horizon
$N_{PartsPerBin}$	Number of parts per bin
O_t	Order quantity in period t
P	Part price
$\mathcal{P}$	Set of parts
p_i	Penalty cost for high order sizes of part i
$q_{0.99,d}(TC)$	99% quantile of the total capacity requirement of demand
$q_{0.99,o}(TC)$	99% quantile of the total capacity requirement of orders
QR	Quantile ratio. $\frac{Q_{99}(Orders)}{Q_{99}(Demand)}$
$\mathcal{R}$	Set of replenishment policies
r_{cv^2}	Ratio of cv^2 of orders divided by cv^2 of demand
r_{Markup}	Markup rate
ρ	Average utilization of the queuing system
ρ_T	Utilization threshold at which $N_{T,max}$ is reached.
ϱ	Correlation coefficient
$\mathcal{S}_{HHR}$	Set of parameter combinations of heavy high runners scenario
$\mathcal{S}_{HLR}$	Set of parameter combinations of heavy low runners scenario
TC	Total required capacity in the leveling horizon
$\widetilde{TC}$	Total chargeable capacity in the leveling horizon
$TC_{d,t}$	Total transport capacity requirement of demand on day t
$TC_{o,t}$	Total transport capacity requirement of orders on day t
$\mathcal{T}$	Set of periods in planning horizon (e.g. days), $\{1, \dots, N_T\}$
t_{Tour}	Tour duration

θ_v	Binary decision variable: Use/do not use vehicle v
$\vec{u}$	Probability distribution of work balance
$\mathcal{V}$	Set of vehicles with different sizes/payloads
v_a	Coefficient of variation of the inter-arrival time or capacity of a continuous-time G|G|1 system
v_b	Coefficient of variation of the service time or demand of a continuous-time G|G|1 system
w_i	Capacity requirement of part i
$W_{ShipmentUnit}$	Weight per shipment unit
W_{Part}	Weight per part
W_{Bin}	Weight of empty bin
x_{it}	Order Quantity of part i on day t
y_{it}	Deviation from mean demand of part i on day t
$\vec{z}$	Probability distribution of deficit
Z_t	Deficit in Period t

1 Introduction

I never predict anything, and I never will.
(Paul Gascoigne, English football player)

Back in the days of Henry Ford, variability was not an issue in production systems. The customer could buy any car he wanted. As long as the color was black, and the model was the Model T, the only model that was available, the customer had the freedom of choice. In this low-variability environment, mass production, as proposed by Taylor and implemented and brought to perfection by Ford, was just the right philosophy of making things. This is why the concept became very popular. Other western manufacturers of automobiles, like General Motors in the early 1930s or Volkswagen in the early 1960s, quickly adapted the concept and participated in the success story of mass production. (Hopp and Spearman 2011, Womack and Jones 1990)

The idea of mass production is to produce every product in high quantities to achieve economies of scale. Since the variety of products is low, each production line can be dedicated to one product or even one product variant, which eliminates the need for setup times. This concept is very capital-intensive, since for each product, dedicated lines from the press shop up to final assembly are required. If the demand is sufficiently high and stable, the concept works great.

Toyota, in contrast, was not able to participate in mass production. The Japanese car industry was still shaken from WWII and did not have the capital resources to build a system of mass production. Even if they had had the capital, they would not have had access to markets where they could have sold their potentially-mass-produced cars. The domestic market in Japan was shaken from WWII as well and thus very small. Because of restricted capital resources and the low demand, Toyota was forced to produce many different product variants on just a few lines. From the beginning on, the Toyota Production System (TPS) had to deal with variability.

In the second half of the 20th century, the market conditions in the market for cars changed. Customers were no longer satisfied with the low variety of products. The car manufacturers responded by extending their product portfolios and increasing the number of models and the number of variants per model. Moreover, the advance of technology and the customer's demand for this technology lead to shorter product life cycles, demanding more flexibility from manufacturers. Although the total demand for cars was still growing, the share per variant decreased because of the increasing number of variants. This caused different problems for the mass producers. It was no longer possible to dedicate whole lines to only one product since the quantities were too low. This led to dramatic decreases in productivity because of the manufacturer's inability to perform quick set-ups. The inflexibility, caused by the new emerging variability, led to waste. (Hopp and Spearman 2011, Womack and Jones 1990, Liker 2004)

Toyota, on the other hand, faced the same change of market conditions. In contrast to the western car manufacturers, they kept their high productivity and were profitable. The reason is that from the beginning on, the Toyota Production System aimed at eliminating the system inhibitors of waste, variability and inflexibility and was designed for producing a high variety of products in small lot sizes and limited capacity. Central elements of the TPS are a culture of continuous improvement and design measures which lead to a "stabilization" of processes. (Womack and Jones 1990, Liker 2004)

Nowadays, the techniques of the Toyota Production are widespread and applied successfully among industry practitioners under the name of Lean Management. In the field of production logistics, many companies have already achieved a high level of maturity. In the field of warehousing, some companies have started introducing initiatives of operational excellence, in which they transfer elements of lean production to the warehousing environment. Prior research suggests that the application of certain elements of lean production in a warehousing environment lead to significant increases in productivity (Dehdari 2013). In contrast to warehousing and production, the penetration of lean management is still very low in transport logistics systems. Milk runs, the ideal concept of transportation according to lean principles, are seldom applied in inbound logistics systems because they are not competitive in terms of costs. One reason for this is – amongst others – the lack of stability in systems of transport logistics which results in a low utilization of the means of transport.

The goal of this thesis is to show how design measures for the stabilization of production logistics systems can be transferred to transport logistics systems. Therefore, we first explore how stability can be defined and how we can measure it. Based on that, we describe how we need to design the system to increase the stability compared to the status quo systems of transport logistics. We investigate how effective the measure is and which factors influence the effectiveness. Moreover, we measure the efficiency by the example of a real-world case study from a German automotive supplier.

1.1 Problem Description and Research Questions

In literature regarding lean production systems, the term "stability" is applied frequently. Different works state that stability is decisive for the success of lean management (cf. Liker and Meier 2006). Up to date, a formal, uniform definition of stability in association with lean production systems is not given. Therefore, in order to create stability in transport logistics systems, we first need to understand what our goal is and establish a proper definition of stability. This results in the following research questions:

> **1st Cluster of Questions:** What existing definitions of stability of logistics systems and related fields are given by literature? What are their similarities and their differences? How can we quantitatively measure stability and stabilization? What is stabilization and how can we achieve it?

Once we have established a definition of stability and a theoretical concept of how we can achieve stabilization, we want to elaborate system design measures to create stability in systems of transport logistics. Therefore, we first need to understand the functionality of status quo systems of transport logistics. Based on this understanding, we can show how a transport logistics system can benefit from stability. Questions regarding transport logistics systems in the status quo are:

> **2nd Cluster of Questions:** What are the elements that transport logistics systems consist of in the status quo? How do they interact? What concepts are applied to fulfill the physical flow of material? What concepts are employed to control the flow of information? How can the flow of material and the flow of information be combined? How can transport logistics systems benefit from stability?

Based on the description of status quo transport logistics systems and their need for stability, we show how transport logistics systems need to be designed in order to achieve stability. Therefore we transfer the production control logic for leveling of lean production systems, "heijunka", to transport logistics. To study the system behavior, we build mathematical models of the system. Moreover, we describe the different steps we need to follow to design the system. In the design process, the mathematical models are employed to calculate the buffer inventory and the leveling pattern.

> **3rd Cluster of Questions:** How can heijunka leveling be transferred from production logistics in order to stabilize a transport logistics system? What mathematical models can be employed to describe the system behavior? How much buffer inventory is necessary if demand is fluctuating but replenishment is leveled? How can we calculate a heijunka pattern? Which steps do we need follow in order to design the supply network?

After the system has been described and necessary steps for the system design have been explained, we want to measure the effectiveness of the leveled replenishment policy in a system of transport logistics. For the investigation we create a simulation model and conduct different simulation experiments on a test data set. These test data enable us to determine influencing factors which affect the effectiveness on part level and on total transport level. This is summarized in the following questions:

> **4th Cluster of Questions:** How can a system with heijunka-leveled material replenishment be modeled in simulation? What is the effect of a leveled replenishment policy regarding the total required transport capacity? What is the effect on the individual part level? What influencing factors affect the effectiveness?

Since heijunka leveling creates stability by moving variability from the capacity dimension to the inventory dimension, the optimum point of operation is characterized by a trade-off: the more capacity we save, the more buffer inventory is required and vice versa. The location of the optimum or efficient point of operation depends on both the factor costs of transport capacity and buffer inventory. It is given by the minimum of the system's operating cost, i.e. the sum of capacity and inventory cost. To determine the location of the optimum, we express the system costs as a function of the buffer allocation. Then, we take its derivative to obtain the marginal cost function and the cost-minimal buffer allocation.

> **5th Cluster of Questions:** What is efficiency and how can we evaluate it for the scope of transport logistics systems? How can the trade-off between inventory and capacity buffers be modeled in terms of a cost function? What is the optimum buffer allocation and how can we calculate it?

All the results and answers that are obtained for question clusters 1 to 5 are based on artificially created demo data and abstract models. In order to ensure that the methods and models that are created in this work are transferrable to industrial practice, we introduce a case study of the Germany Automotive Supplier ZF Friedrichshafen as a numerical example. We first provide some descriptive analyses of the data that is used for our case study. Afterwards, we present how to process the data for our proposed Design for Stability. We calculate the efficient point of operation according to the methods developed in this work. Moreover, we compare the system we designed to a status quo transport logistics system.

> **6th Cluster of Questions:** How can the methods and models that were developed in this work be applied to design a real-world example? What is the cost-minimal buffer allocation according to the different models? What are the system's operating costs in comparison to a status quo transport logistics system?

1.2 Structure of the Thesis

Based on the research questions and problem statement described in the preceding section, we will explain the structure of the thesis in this section. Figure 1.1 depicts a summary of our thesis outline.

The second chapter aims at establishing a definition of stability which is applicable in the field of transport logistics. We review different existing definitions of stability which originate from different fields of applications and derive a definition which is employed in this thesis. Moreover, the term stabilization is defined and a specification of how stabilization can be achieved in logistics systems is given.

The third chapter lays the theoretical groundwork of this dissertation by explaining the basics of transport logistics systems. The first section reviews the physical flow of material by presenting different transport concepts which are applied in the German automotive industry. The second section analyzes the flow of information by presenting different control policies which are applied in materials supply. The chapter closes with the synthesis of the flow of material and the flow of information. In this synthesis, we point out the need for stabilization in transport logistics and present the benefits that can be leveraged by an increase in stability.

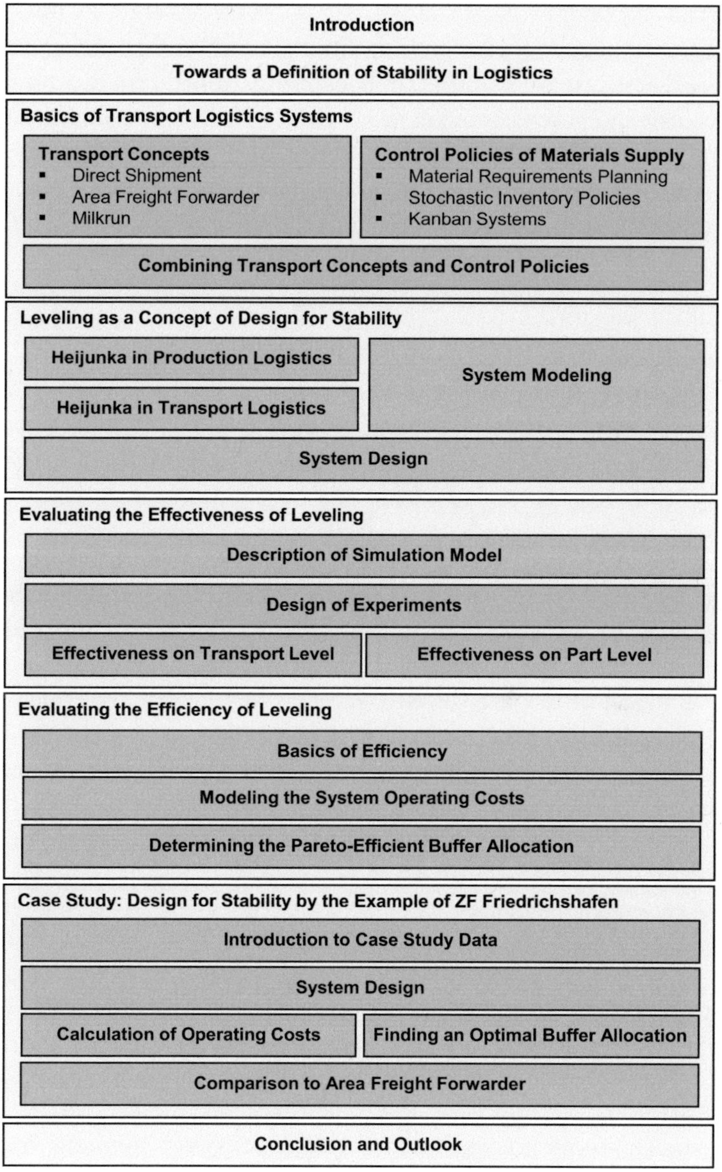

Figure 1.1: Structure of this thesis

The fourth chapter provides a concept of "Design for Stability" by transferring the control strategy heijunka leveling from production logistics to transport logistics. Therefore, we explain the functionality of heijunka in a production environment and how the system needs to be modified to stabilize transport logistics systems. To study the system behavior, mathematical models of the system are presented. These models are then employed to design the system, i.e. to calculate a heijunka schedule and the resulting necessary buffer stock.

The fifth chapter evaluates the effectiveness of leveling as a concept of "Design for Stability". An agent-based simulation model which is able to evaluate the measure effectiveness on individual part level and on transport level is proposed. A set of test data which incorporates the Pareto distributions of both unit weight and mean demand is presented. Based on these test data, we conduct a simulation study to determine the effectiveness of heijunka leveling with respect to stabilizing the required transport capacity and investigate effectiveness-influencing factors.

Chapter 6 evaluates heijunka-leveled transport systems in terms of efficiency. First, the basics of efficiency in the sense of Pareto are explained and transferred to the scope of transport logistics systems. Afterwards, we build a universal model of heijunka-leveled transport logistics system which allows for the computation of the system's total operating cost as a function of the buffer allocation. Based on this cost function, we derive the marginal cost function and determine the Pareto-efficient buffer allocation as a cost-minimal point of operation. Moreover, we show how the location of this optimum changes by enumerating over different factor cost ratios. In addition, we develop a simple analytical formula which allows a quick approximate computation of the optimum catch-up capacity.

Chapter 7 presents how the methods presented as part of our "Design for Stability" can be applied to a real world example as a special case of the abstract model presented in the sixth chapter. In a case study which is based on data of the German Automotive Supplier ZF Friedrichshafen, we design the system by allocating transport concepts, forming tours, generating heijunka patterns and calculating necessary buffer stocks. We determine the Pareto-efficient buffer allocation according to different models and compare the results to evaluate their accuracy. Furthermore, we compare the operations costs of the heijunka controlled system to the status quo system which is controlled by Material Requirements Planning (MRP) in association with an area freight forwarder.

Chapter 8 summarizes the major findings and contributions of this work and draws a conclusion. The thesis closes with an outlook which provides directions for further research.

2 Towards a Definition of Stability in Logistics

If something's hard to do, then it's not worth doing.
(Homer Simpson, TV character)

This chapter aims at establishing a definition of stability for logistics. The first part reviews the terms "stability" and "stabilization" from a linguistic point of view by describing their definition according to the dictionary. Afterwards, we review existing definitions of stability in the field of logistics and relate them to the linguistic definition. The third part investigates how stability is defined in in statistical process control and how it can be quantified. The fourth part further introduces the concept of variability in association with production systems and explores its relation to stability. The chapter closes with a conclusion in which we summarize our understanding of stability in this work.

2.1 Linguistic Use of the Term Stability

The Oxford Dictionary of English defines stability, from Latin stabilitas, as the state of being stable (Oxford 2012). Looking up the term "stable" yields the following entries:

stable - adjective

- *(of an object or structure) not likely to give way or overturn; firmly fixed: "specially designed dinghies that are very stable"*
- *(of a patient or their medical condition) not deteriorating in health after an injury or operation: "he is now in a stable condition in hospital"*
- *Sane and sensible; not easily upset or disturbed: "the officer concerned is mentally and emotionally stable"*

- *Not **likely** to **change or fail**; firmly established: "a stable relationship", "prices have remained relatively stable"*

- *Not liable to undergo chemical decomposition, radioactive decay, or other physical change: "isocyanic acid reacts with amino groups to form a stable compound", "stable nuclei"*

Taking a closer look at the meaning "not likely to change or fail" we identify the following elements:

- **Likelihood** and

- to fail, i.e. **failure**.

Likelihood, according to Oxford (2012), is synonymous to probability. Looking up "to fail" yields the definition "to be unsuccessful in achieving one's goal". A goal is an aim or a desired result. Therefore we need some kind target state which defines whether our goal was reached or not, i.e. our attempt in reaching our goal was successful. The expression "not likely to change or fail" can therefore be understood as a low probability of the outcome of a process being out of a desired target state.

Closely related to stability is the term stabilization. Looking up the term in the Oxford (2012) yields the following entry:

stabilization - noun

- *The process of making something physically more secure or stable. "the derelict buildings will require some structural stabilization"*

- *The process of becoming or being made unlikely to change, fail, or decline. "the economy is starting to show signs of stabilization", "stabilization of the patient's cardiac function", as modifier "the corporation's stabilization fund was still in arrears"*

This definition fits with the preceding definition of stability. Therefore, we conclude to stabilize means to increase the probability of a process outcome being within a desired target state. This equivalent to reducing the probability of failing.

To conclude the preceding discussion we can state that, in order to describe the stability of a process, we need a **target state** for the process outcome and a **probability** of the process outcome being within this target state. Stabilization is the **increase of the probability** of a process outcome being within a target state.

2.2 Existing Definitions of Stability in Logistics

In order to find a definition for stability in logistics, we also would like to review some existing definitions of stability in fields which are related to logistics. Therefore, we first describe the definition of Daganzo (2003) who elaborates a control theoretical understanding of stability of supply chains. Afterwards, we review the dissertation of Meissner (2008) who analyzed the stability of mixed flow lines in the automotive industry.

2.2.1 Logistic Stability in Automotive Mixed Flow Lines

Another attempt of finding a definition for logistics can be found in the dissertation about the sequence stability in automotive mixed flow assembly lines of Meissner (2008). In his Dissertation, Meissner provides approaches and methods for stabilizing the order sequence of automotive flow lines. To evaluate these approaches and methods, Meissner first describes what he means by the term stability and later provides performance figures to quantify stability.

Similar to Daganzo (2003), Meissner (2008) also derives his definition of stability from control theory. According to the control-theoretical definition of stability, a stable system is able to return to a desired state after a disturbance.

That is, as operational disturbances frequently occur in production systems, the performance variables can temporarily deviate from the target value. If the system is stable, these performance variables gradually return to their target value. This is displayed for a disturbance at time=0 in Figure 2.1

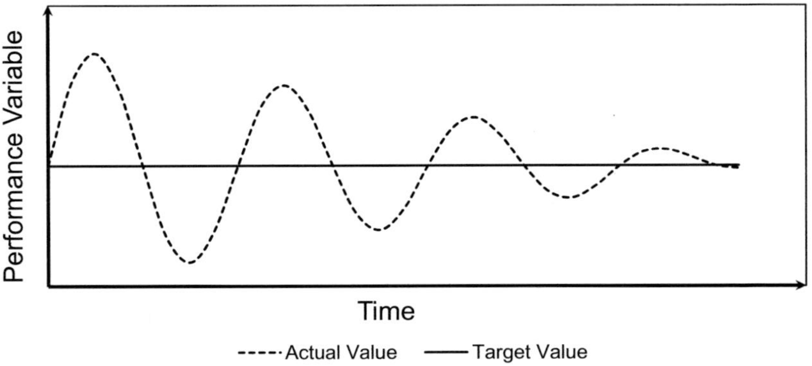

Figure 2.1: Stable variable as a function of time after a disruption at time=0 (adapted from Meissner 2008)

Based on this control-theoretical model of production systems, Meissner states that stability in the flow of production is the minimization of the dispersion of actual process performance around its target value. During the time of the disruption, the requirements of the subsequent process or customer are fulfilled at any time. In a stable system, short- term disruptions still exist. That is, stability is not the absence of disruptions. The stable system is characterized by the fact that process performance always recovers to the target value. From this definition, we can derive that instability must be a kind of system state where the dispersion of process performance around the target value is so high that the customer's demand cannot be satisfied.

According to Meissner, two central objectives for the improvement of processes in production logistics which need to be distinguished are the shortening and stabilization of lead time. Whereas lead time shortening aims at reducing the mean or expected value of the lead time, stability points at reducing the dispersion of lead time around today's mean lead time $\overline{LT}$, as depicted in Figure 2.2.

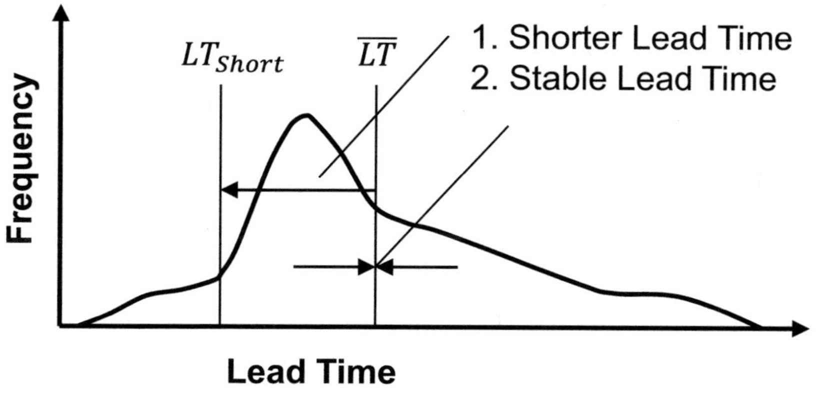

Figure 2.2: Stability in Logistics according to Meissner (2008)

To quantitatively measure and evaluate the methods he provides for the stabilization of the order sequence, Meissner proposes several performance figures:

- Adherence to delivery dates [%]
- Standard deviation of lead time
- Mean sequence deviation
- Standard deviation of sequence deviation

Meissner's definition of stability concurs with the definition from Oxford (2012) given in section 2.1 regarding the point that there must be a certain target state. This state is described as a single value, i.e. not a range of values. In contrast to section 2.1, Meissner (2008) does not explicitly take into account probabilities and does not distinguish between probabilities of failure and success.

Moreover, Meissner's definition does not differentiate between different intensities of stability. That is, according to the definition a system can be stable or instable. The definition does not allow for two system that are both stable, but one of these systems is more stable. Stabilization therefore can only incur if a system initially is instable. A stable system cannot be stabilized any further.

2.2.2 Supply Chain Stability

Daganzo (2003) discusses the term stability not looking at individual processes but at the supply chain as a whole system consisting of multiple tiers. Focus of the investigation is the supply chain behavior under stochastic demand.

The model of Daganzo (2003) is depicted in Figure 2.3. The supply chain system is triggered by the customer orders $N_0(t)$. The state of each supply chain tier i is described by the inventory position $x_i(t)$ and the physical inventory $y_i(t)$. The inventory position is defined as the sum of physical inventory and open orders. According to an order policy (see section 3.2.1), orders $u_i(t)$ are placed at the upstream entity. The order quantity depends on historic orders of the direct downstream entity u_{i-1}, i.e. the consumption of the direct customer.

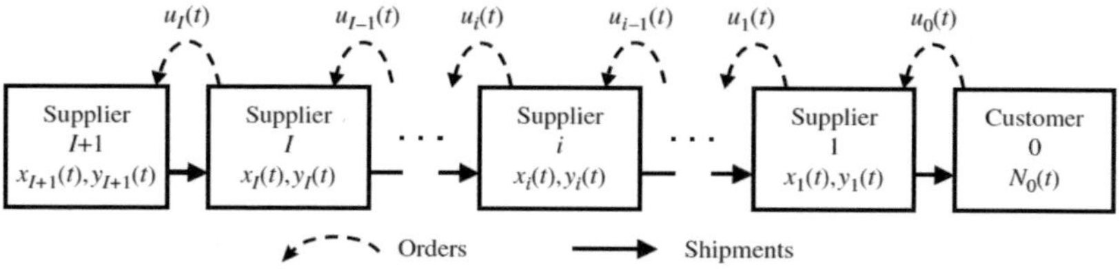

Figure 2.3: Supply chain model of Ouyang and Daganzo (2006)

A supply chain is called instable if the variability of the order quantity or inventories increases when we move upstream, starting at the customer. That is, we can observe the so-called bullwhip effect[1]. Put differently, a supply chain is stable if it does not encounter the bullwhip effect.

Daganzo (2003) distinguishes between stability in the small and stability in the large. A supply chain is said to be stable in the small, if an infinitesimally small deviation at the input leads to a bounded (small) deviation at all entities of the supply chain, even if the supply chain is infinitely long. This implies, that at any supply chain entity, we can reduce the deviation from the steady state (i.e. all entities place constant orders) down to a desired bound by decreasing the steady-state deviation of the input signal. This is for example not the case, if minimum order quantities increase as we move upstream the supply chain. In this case, a small deviation at the downstream process either leads to no deviation or a substantially larger deviation at the upstream process. Another implication of this stability in the small is that a small change at the customer leads to small changes at all entities along the supply chain. Therefore, a policy of "if a change of demand occurs, do not change anything" is not permitted.

In contrast, a supply chain is said to be stable in the large if a very large (but not infinite) deviation from the steady state at the input leads to bounded orders at all stages of the supply chain, even if the chain is of infinite length.

[1] The bullwhip effect describes the amplification of demand oscillations along the supply chain. It was named for the way the amplitude of a whip increases down its length.

If the conditions both for stability in the small and stability are satisfied, the supply chain is said to be strongly stable.

Note that Daganzo's definition of stability also has a target state, as described in section 2.1. Both the stability in the small and in the large define bounds which must not be exceeded in order to achieve stability, because otherwise the bullwhip-effect will be present in the supply chain. However, the approach is purely deterministic and the definition of stability does not take into account probabilities and it does not provide a way of achieving stabilization.

2.3 Stability in Statistical Process Control

Statistical Process Control is a method of quality control which uses statistical methods to monitor and control processes, usually in a manufacturing environment. The central goal is the management and reduction of variability in production processes in order to achieve stable outcomes. More precisely, this means the processes must be capable of operating with little variation around the target dimensions of the products' quality characteristics. (Montgomery 2009)

In a manufacturing environment, in any process, regardless of how well designed or carefully maintained, there will always be a small amount of so-called "natural variability". That is, the outcome of a process never is exactly the same, but there is some kind of variation in it. It is the cumulative effect of many small unavoidable causes. If this variation is within certain specified limits and the system is operated only with natural variability, i.e. unavoidable and unassignable causes, we speak of a "stable system of chance causes". That is, the system is in statistical control. The term chance causes implies that there still is some variation in the process, but it only happens due to chance and not due to an assignable cause.

Besides natural variability, it is possible that there is another kind of variation present in the output of a process, which is called variability caused by an assignable cause. This kind of variability is usually large compared to natural variability. If an assignable cause is present in a process, it is called an out-of-control process and the system is not considered as stable.

Natural variability and variability caused by an assignable cause can be explained by the example of a forming press which changes the forms of parts. If the die is new and the press works fine, the parts that are shaped are never perfectly identical and there still is some variation in each shaped part. If the variation is random and within a certain tolerance, the parts are fit for use. When the die wears down, the variation is no longer within the range that is tolerated. Since the reason for the variation is the wear down of the die, there is an assignable cause.

Figure 2.4 illustrates the impact of assignable and chance causes on an arbitrary process quality characteristic. Speaking in terms of stability, we would call the processes instable (or non-stable) and stable. Until time t_1, the process in the figure is stable. Both mean and standard deviation are at their in-control values.

Starting at time t_1, an assignable cause occurs. This causes the mean to shift from μ_0 to $\mu_1 > \mu_0$. At time t_2 a different assignable cause occurs. Now the mean is $\mu = \mu_0$, but the process standard deviation has shifted to a larger value $\sigma_1 > \sigma_0$. At time t_3, another different assignable cause occurs which shifts both process mean and standard deviation. For all t_i with $i > 0$, the process is not stable, i.e. out of control.

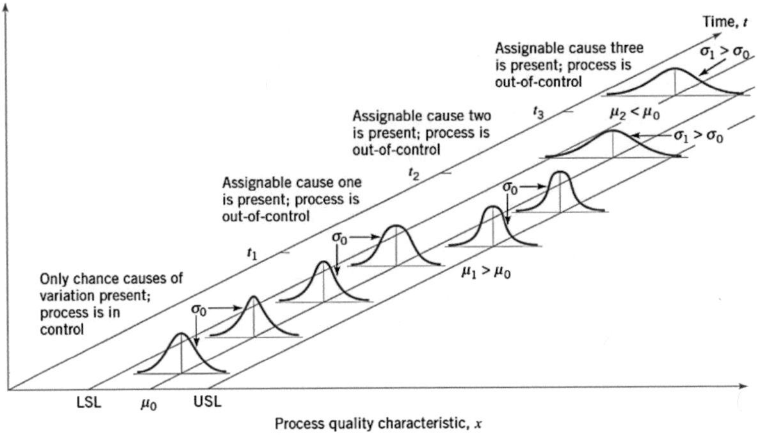

Figure 2.4: Chance and assignable causes of variation (Montgomery 2009)

Note that in Figure 2.4, there are the upper (USL) and lower (LSL) specification limits. When the process is stable (i.e. in control), most of the probability mass of the displayed distribution lies within these specification limits. When the process is out of control, the amount of probability mass within these limits decreases. Therefore, the range within these specification limits can be interpreted as a target state, as described in section 2.1.

Moreover, as described in section 2.1, a proper definition of stability needs a target state and a likelihood of a process outcome being inside or out of this target state. Therefore, we now define the stability at an α-level as:

- $P(LS \leq X \leq USL) = (1 - \alpha)$
- $P(X<LSL) + P(X>USL) = \alpha$

This means, the probability of our process outcome X being within our target state, as specified by the specification limits, is $1-\alpha$. Vice versa the probability of the process outcome being outside the target state, is α.

This definition concurs with the definition of Meissner (2008), as given in section 2.2.1 and extends it with further details. In contrast to Meissner, the target state in statistical process control is defined as a range of values. In mathematical statistics, although not explicitly stated by Meissner (2008), the concept of "minimizing the dispersion around a target value" is equivalent to "maximizing the probability of the process outcome being within a desired range".

The approach to defining stability which is yielded by statistical quality control allows both for an incorporation of a target state and the likelihood of a process outcome being within this target state. Therefore it also provides a fit with the linguistic definition from the dictionary, which we provided in section 2.1.

2.4 Stability, Variability and Stabilization

Hopp and Spearman (2011) provide some useful insights regarding stability with their analysis of variability in production systems. Variability is an inherent part in all production systems and usually cannot be eliminated completely. Variability degrades the production systems' performance. Therefore the goal is to reduce variability as much as possible.

The authors distinguish between controllable variation, which is the result of decisions, e.g. offering different product variants, and random variation, which is a consequence of events beyond control, e.g. varying customer demand (c.f. section 2.3). Controllable variation can be influenced by improvement processes, e.g. by introducing standard work and a process of continuous improvement. Random variation is exogenous to the production system and therefore it cannot be influenced directly or only with a certain effort. Varying customer demand, for example, is often a result of the customer's overall equipment effectiveness (OEE). The reason is that the customer usually demands the parts he has consumed before. This depends on how many products he was able to make. In general, it is not possible to directly change the customer's OEE, therefore is regarded as exogenous here[2].

As a quantitative figure to measure the variability of a random variable X, the authors provide the squared coefficient of variation, which puts the variance of a random variable in relation to its expected value. If cv^2 is equal to zero, the process is deterministic. If it is equal to one, the process is random.

$$cv^2 = \frac{Var(X)}{E^2(X)} \tag{2.1}$$

The authors classify different sources of variability in a production system:

- Natural variability
- Preemptive random outages, i.e. unplanned outages or breakdowns
- Variability from non-preemptive or planned outages, e.g. setups or preventive maintenance
- Variability from recycle, e.g. rework due to quality issues.

Natural variability is a catchall-category which encompasses the variability inherent in natural process time. It accounts for variability from sources which have not been explicitly called out, like e.g. a piece of dust in the operator's eye. Many of these unidentified sources of variability are operator related. This is why there is typically more natural variability in manual production processes than in automated ones. External influences like random downtimes or setups are not included in this category.

The second source of variability are so-called preemptive outages. These are unscheduled downtimes which increase both the mean and cv^2 of effective

[2] Of course there are exceptions to this claim. If the customer belongs to the same company as the producing plant, it is likely that both cooperate on improving the OEE. Moreover, it also possible that supplier and customer cooperate as part of an improvement process, as for example Toyota and its suppliers (see Womack and Jones 1990).

process time. In many production systems, this is the single largest cause of variability. An example of preemptive outages are breakdowns. These occur whether we want them or not and at any time, e.g. right in the middle of a job. Other possible sources of preemptive outages are power outages, operators being called away on emergencies or running out of consumables.

In contrast to pre-emptive outages, non-preemptive outages represent downtimes which inevitably occur but we have the possibility to control when. An example of a non-preemptive outage is wear and tear of a machine tool which requires a replacement. In these cases, the current job which is in process can be completed before we need to stop the machine to replace the tool. Setup times, since they are usually planned and scheduled, and preventive maintenance are further examples of non-preemptive outages.

Variability from recycle originates in quality problems from poor processes in the production system. If we think of a single machine or workstation which performs a certain job. Later during quality check we note that the job was not performed properly. Therefore, we need to repeat the task to get the job right. This leads to additional processing time for this workstation. If there is a quality check after a chain of different processes and the mistake was at the beginning, this might lead to additional process time for all machines which belong to the process chain. If we interpret the additional processing time as an outage, we note that this kind of variability is similar to the non-preemptive outage case.

Further, the authors propose the so-called 'Law of Variability Buffering'. According to this law, variability, caused by either of the above sources, must be buffered by some combination of capacity, inventory or time. Buffering can be realized by buffering in the same dimension, e.g. time buffers for time variability, or by shifting variability by control mechanisms from one dimension to the other. If the variability is controllable, we can reduce it by process improvements which results in less buffering.[3] If variability is random

[3] According to Hopp and Spearman (2004), 'Production of goods or services is lean if it is accomplished with minimal buffering costs'. Therefore the goal of lean is to reduce variability as much as possible, which reduces the necessary buffers.

or exogenous, the only way to satisfy the customer's requirement is to allocate buffers.

An example of buffering in the same dimension is to start early for a trip with variable travel time in order to arrive on time. In this case, we allocate a time buffer to account for the variable travel time to ensure we arrive before a specified time.

An example of shifting variability from one dimension to another by control mechanisms is the heijunka leveling technique from the Toyota Production System. Heijunka leveling, in the make-to-stock variant, is a control mechanism which decouples the production system from varying customer demand by an inventory buffer. Since the requested quantities vary from day to day, we would need a varying production capacity from day to day. Because production capacity depends on manpower and machines and is thus rather fixed, it cannot be varied synchronous to customer demand on a short-term basis. One option would be to buffer the varying customer demand by a capacity buffer. That is, we would increase our production capacity that much that even demand peaks can be satisfied. This would result in poor capacity utilization on days on which there is no peak in demand. Since manpower and machines are expensive, this alternative is very cost-intensive. In heijunka leveling, we reserve a production capacity which is able to fulfill the mean customer demand and buffer the variations by the aforementioned inventory buffer. In cases where holding inventory is cheaper than excess production capacity, it makes sense to take this decision.[4]

In the case of make-to-order leveling, the customer must be willing to accept a certain lead time between the placement and the arrival of the order. Therefore, the producer is able to smooth the orders over time, i.e. the orders of two succeeding days of too high and too low utilization are consolidated to utilize the production system just right. This is displayed in Figure 2.5.

[4] These are not the only advantages which originate from introducing heijunka leveling to a production system. Heijunka is also creating transparency in the production system and is an important tool for driving the continuous improvement process.

In the example, our production capacity each day is 5 units. Since our customer informs us in advance about the quantity he likes to order, we know he will need 2 units on Tuesday and 8 units on Wednesday. Producing exactly what the customer wants in not possible, since our capacity is only 5 units. Thus, we smooth the production quantity by producing 5 units on Tuesday as well as on Wednesday.

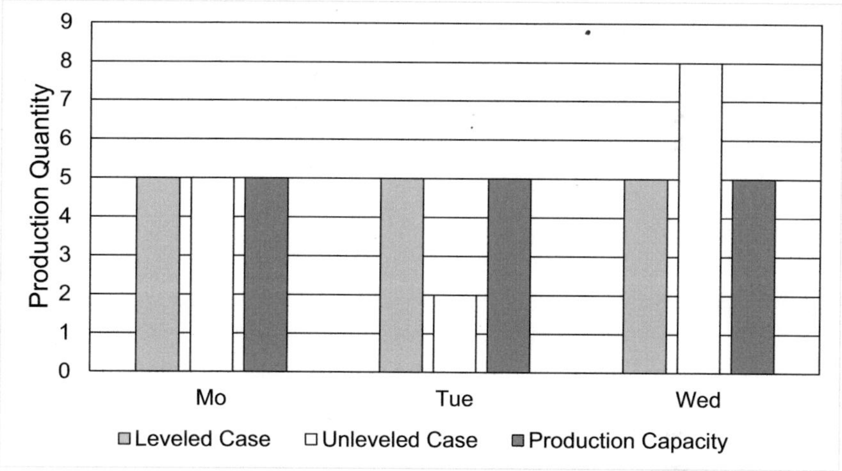

Figure 2.5: The idea of make-to-order leveling

If the customer would need 8 units on Tuesday and 2 units on Wednesday, we would not be able to fulfill the demand. That is why the customer must be willing to accept a certain order lead time. This way we can move the quantities on days of above-average demand along the time axis to days of below-average demand, i.e. days on which there is more capacity than demand. That is, the goods are produced earlier than they are needed by the customer and put into stock. The shipment would be performed just before the goods are needed by the customer.

In contrast to Daganzo (2003), Meissner (2008) and Montgomery (2009), Hopp and Spearman (2001) do not explicitly address the term stability. However, their remarks regarding variability in production are worthwile for our investigation of stability. The coefficient of variation, which is proposed by Hopp and Spearman as quantitative figure of variability, is a measure of dispersion around a target value, as described by Meissner (2008). Combining the works of Meissner and Hopp and Spearman, yields the conclusion that for the achievement of stability, we need to minimize variability. According to Montgomery (2009), we can measure stability by the probability mass of a random variable within specified limits. Reducing the variability of this random variability is equivalent to increasing the probability mass inside these limits. Thus, variability and stability can be understood as antagonists, since increasing variability reduces stability and vice versa. Therefore, the three different approaches fit with each other and do not contradict.

The Law of Variability Buffering yields further insights regarding stability and stabilization. Assuming that a stable system is characterized by the fact that buffers are allocated properly so that customer requirements can be fulfilled, we can categorize three kinds of measures to achieve stabilization:

- Buffering in the same dimension (e.g. time buffer for time variability)
- Move variability into another buffering dimension (e.g. use buffer inventory for the stabilization of capacity)
- Directly decrease process variability (e.g. introduce standard work, preventive maintenance)

Each of these measures of 'stabilization' results enable us to take an influence on the probability mass of a certain process outcome (in terms of capacity, inventory or time) being within a desired target range.

2.5 Conclusion

In this section we reviewed different works which take an attempt at defining stability. We found that, according to the dictionary, stability can be defined as 'unlikely to change or fail' which equivalent to 'likely not to change or not to fail'. Failing means being unsuccessful in achieving one's goals. Therefore, we concluded the stability is a high probability of a process outcome being within a desired target range.

Based on this linguistic definition, we reviewed works related to stability in different fields which are related to logistics. These are summarized in Table 2.1.

Daganzo (2003) builds a control-theoretical of a supply chain and investigates the amplification of order quantities along the supply chain. He states that, if the amplification factor is greater than 1, the supply chain is subject to the bullwhip-effect and the system is not stable. His approach is purely deterministic and does not take into account probabilities. Therefore, the approach is in conflict with definition given by the dictionary.

According to Meissner (2008), the production flow is stable if the system, despite short-term operational disturbances, recurs to its target state while fulfilling customer requirements. This state can be measured quantitatively if the dispersion of process performance around the target value is minimal. Meissner does not state an absolute value which yields a conclusion to which extent the dispersion needs to be minimized. Therefore, the stability of a system can only be compared relative to another system, i.e. not on an absolute scale. Since Meissner's approach is stochastic and envisions a target state, the definition concurs with the linguistic definition.

We note that Meissner states that a stable system is characterized by satisfying customer requirements while there is a varying (dispersing) process performance. The statement implies that there must be some kind of reserve to buffer the varying performance. This fits with the elaborations of Hopp and Spearman (2011), as elaborated in section 2.4.

Table 2.1: Literature overview – Stability in Logistics

	Daganzo (2003)	Meissner (2008)	Montgomery (2009)	Hopp / Spearman (2001)
Application Field	Supply Chains	Mixed Assembly Flow Lines	Quality Control	Production Systems
Definition of stable state	No Bullwhip-Effect in Supply Chain	Minimal dispersion around target value and customer satisfaction	Process outcome with desired probability within specification limits	Fulfilling the customer requirements
Approach	deterministic	stochastic	stochastic	stochastic
Definition of Target State	amplification < 1	low dispersion around single target value	within absolute specification limit	not explicitly given
Control Variable	replenishment orders	order sequence	quality characteristic (generic)	inventory, capacity, time
Indicator Variable	amplification factor	adherence to delivery date, standard deviation of lead time	probability mass in specification limits	cv^2
Scale	relative	relative	absolute	relative

Montgomery (2009) describes how the term stability is defined in quality control. For an arbitrary quality characteristic, certain specification limits are defined as target state. If a sufficient proportion of the probability mass of the sample measures is within these specification limits, the process is said to be stable. This definition fits with the linguistic definition and matches and extends Meissner's definition.

Hopp and Spearman (2011) do not explicitly address stability but analyze its antagonist, variability. They state that variability must be buffered by some kind of capacity, inventory or time. Insufficient buffers degrade the production systems' performance and prevent us from satisfying the customer's demand. We conclude that a stable system is characterized by properly dimensioned buffers. Stabilization can be thus be achieved by buffering or improvement measures which result in reduction of variability.

In this work we define stability as the **probability of a process outcome being within a desired target state**. The **dimension** of the process outcome can be either **capacity, inventory or time**. The target state must be defined as a range in either of these three dimensions. Stabilization can be measured both by an **increase of the probability mass being within the target state** and by a decrease of the cv^2. Ways to achieve stabilization are to directly **decrease process-inherent variability** or to introduce **buffers** in either the same dimension or by shifting variability from one dimension to another.

3 Basics of Transport Logistics Systems

Facts are meaningless. You could use facts to prove anything that's even remotely true.
(Homer Simpson, TV character)

In this chapter, we elaborate the basics which lay the groundwork for following parts of this thesis. A logistics systems consists of two subsystems: The physical flow of material and the flow of information. In the first section of this chapter we review the physical flow of material by giving an overview of the transport concepts which are currently applied in the German automobile industry. Afterwards, we describe how the physical flow of goods is controlled by the flow of information. That is, we review different control policies which are applied in material supply. These control policies define how orders are triggered and supply is planned. The third part of this chapter is dedicated to their synthesis, i.e. how transport concepts and control policies are typically combined to form a system of transport logistics.

3.1 Transport Concepts of the German Automobile Industry

In this section, we would like to introduce the road transport concepts which are used in the German automobile industry. They are provided by VDA guideline 2010 "Richtlinie Standardbelieferungsformen". (VDA 2008)

Table 3.1 gives and overview of the different transport concepts which are considered in VDA 5010. The first feature by which we can distinguish the different concepts is the shipment size. That is, whether the shipment is assigned a full truck or if the shipment shares the truck with other shipments. In the first case, we speak of full truck load shipments (FTL), in the latter case, we speak of less than truck load (LTL) shipments. That is, if one truck is used exclusively for one shipment, we speak of an FTL shipment. If two shipments are consolidated for one truck, we refer to both of these shipments as LTL shipments.

Table 3.1: Overview of Road Transport Concepts in the German automobile Industry according to VDA 5010

Road Transport Concepts		
Full Truck Load (FTL)	Less than Truck Load (LTL)	
Direct Shipment	Groupage Service	Milk Run

In a full truck load shipment, the shipment size of one supplier is large enough to achieve a sufficient vehicle utilization. Therefore, the transport task is to bring the shipment from the supplier to its destination. In the case of less than truck load shipments, the shipments are smaller. That is, dedicating at whole truck to the shipment would result in a poor vehicle utilization and thus also high costs. This is why the logistics service provider either performs a temporal or spatial consolidation of shipments of multiple suppliers to form larger loads. Since the costs which arise for the Logistics Service Provider (LSP) are more or less fixed, increasing the number of shipments which are being transported leverages economies of scale. Courier- and Express (CEP) service providers, another concept which is employed for sending LTL shipments, are not considered in VDA 5010 since they only play a minor role in the German automobile Industry.

In the following sections, we describe the three transport concepts of Table 3.1, i.e. direct shipment, groupage service and milk run.

3.1.1 Direct Shipment

If we speak of a direct shipment there is a direct point-to-point relationship between the supplier and the receiving plant, as depicted in Figure 3.1. In that case, the receiving plant, or consignee, contracts a logistics service provider to pick up parts at one supplier and deliver them directly to its destination. In between, there is no sorting, unloading or consolidation process (e.g. cross-docking). (VDA 5010)

The concept is best used in association with stable and high demand which is able to form full truck loads at an acceptable delivery frequency. If the consumption is too low, direct shipment becomes uneconomical. Either the delivery frequencies become very low and thus the inventory coverage and hence inventory cost rise, or the truck is not well utilized and a lot of "air" is transported. This is why in these cases, we would not send these shipments as a full truck load but consolidate them with other shipments to form bigger loads.

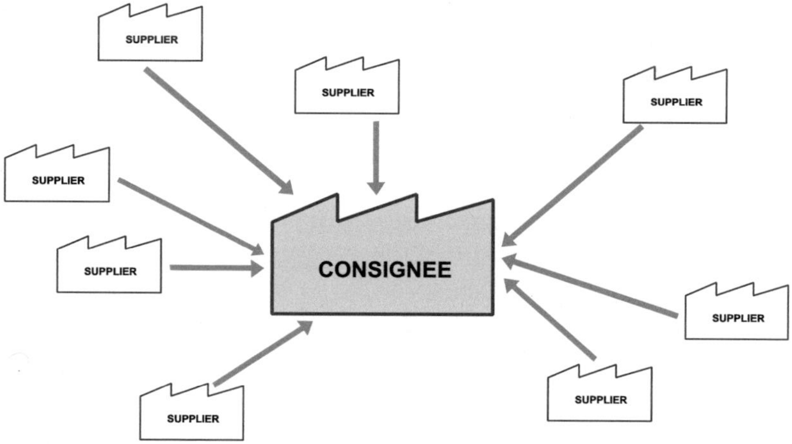

Figure 3.1: Direct shipment with point-to-point relationship between supplier and receiving plant

According to Meyer (2015), the cost for direct FTL transports mainly depends on the driving distance, tour duration and the freight forwarders market situation in the source or destination area, but not on the volume or weight of the shipment. That is, the receiving plant pays for the whole truck capacity, regardless of the actual weight or volume carried by the truck.

As a recommendation by the VDA guideline 2010, direct shipments are best applied in case of regular delivery frequencies and with full truck loads (as already mentioned above). Often, this is associated with Just-in-Sequence (JiS), Just-In-Time (JiT) or VMI delivery concepts. (VDA 2008)

3.1.2 Groupage Service

If delivery frequencies are not sufficiently regular or the loads which need to be transported are too small to fill up a whole truck, direct shipments become uneconomical. In this case, we consolidate several LTL shipments with the goal of creating full truck load shipments.

One concept of bundling several shipments in order to increase vehicle utilization which is widely applied in the German automobile industry is the groupage service which is often offered by so called area freight forwarders (AFF). These are usually logistics service providers or coalitions of smaller freight forwarding companies which are contracted for two to three years to provide transport capacity in certain areas at fixed tariffs. (Meyer 2015)

The term area freight forwarder derives from the fact that in German automobile industry, the network of suppliers is usually divided into several consolidation areas. The criterion according to which the network is divided can be post codes, federal states or countries, for instance. In each of these consolidation areas, a freight forwarder is responsible for planning and executing all the transports which are related to suppliers in their respective area, i.e. mainly planning and executing cost efficient tours. This is convenient for the receiving plant, since all the planning task are outsourced to the logistics service provider. (Grunewald 2014)

Typically, the freight forwarders operate a hub-and-spoke network, which consists of a pre-leg, a main leg, and sometimes, but not necessarily, a subsequent leg (c.f. Figure 3.2). In the pre-leg, the forwarder collects material from all supplier plant surrounding the respective hub in the area. The parcels which are collected at various suppliers do not necessarily all have the same destination. It is rather common that the freight forwarder collects parcels which are all destined for different receiving plants. All the parcels are brought to the forwarder's hub. In the hub, the parcels from various tours, each one consisting of one or multiple suppliers, are consolidated to bigger loads and sorted according to their respective destination. (Conze 2014)

In the main leg, the parcels are either brought to another hub, which is close to their final destination, or directly to the final destination. In case of the parcels being brought to another hub, the parcels are deconsolidated, sorted again and then brought to their destinations.

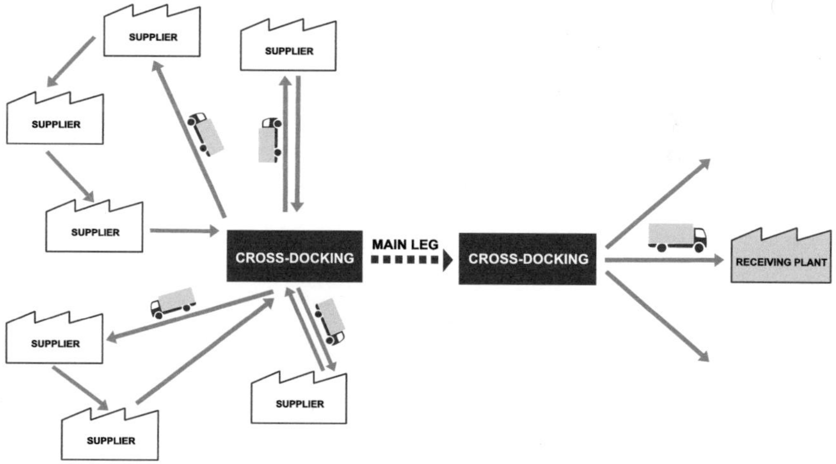

Figure 3.2: Groupage service by area freight forwarders

If the shipping volumes of one or two nearby suppliers already fill up the capacity of a vehicle on the pre-leg, the logistics provider skips the consolidation process and conducts a direct transport from the supplier to the consignee (cf. Schöneberg et al. 2010, Meyer 2015).

The transport capacity which is required by the freight forwarder to fulfill his transport orders varies from day to day. In order to limit this variability, the transport capacity which may be used by one contractor in a certain period of time is contractually limited.

In order to be able to fulfill the demand at an acceptable service level, he must provide a sufficiently large number of vehicles in his fleet. If, however, one day his capacity is too low to fulfill his orders, the freight forwarder subcontracts other freight forwarders to extend his capacity. The higher the variability of the total demand for transport capacity, the lower the average utilization of his fleet.

The variability of the demand for transport capacity is reduced by the pooling effect. The pooling effect is a natural leveling phenomenon and describes the observation that the variance of a sum of random variables can be calculated by adding the variances plus twice their covariance (see equation 3.1). That is, if the random variables are uncorrelated, i.e. their correlation coefficient ϱ becomes zero ($\varrho = 0$), the variations of the two random variables even out. If the demands are positively correlated ($\varrho > 1$), there is an anti-pooling effect. That is, the variation of the sum of the two random variables is greater than the sum of the variations of the two single random variables.

$$Var(X_1 + X_2) = Var(X_1) + Var(X_2) + 2\varrho\sqrt{Var(X_1) \cdot Var(X_2)} \tag{3.1}$$

If the freight forwarder receives a sufficiently high number of orders from various consignees from different industry sectors, however, we can assume that these order are statistically independent, i.e. uncorrelated, and variations even out. Since the freight forwarder has a contractual agreement with the receiving plant which urges him to provide transport capacity which is sufficient to achieve a certain service level, he needs to hold available a respective fleet. The more pronounced the pooling effect, the better is the utilization of the freight forwarder's fleet.

The price that is charged is usually calculated by using a so called tariff table. These contain prices for certain weight and distance classes. Usually, there is a degression in weight. The prices are independent from the actual physical tour. Even if the weight is picked up from multiple suppliers on one tour and then brought to the receiving plant, the distance used for calculating the price is the sum of the distances between each individual supplier and the receiving plant (cf. Figure 3.3). (Meyer 2015)

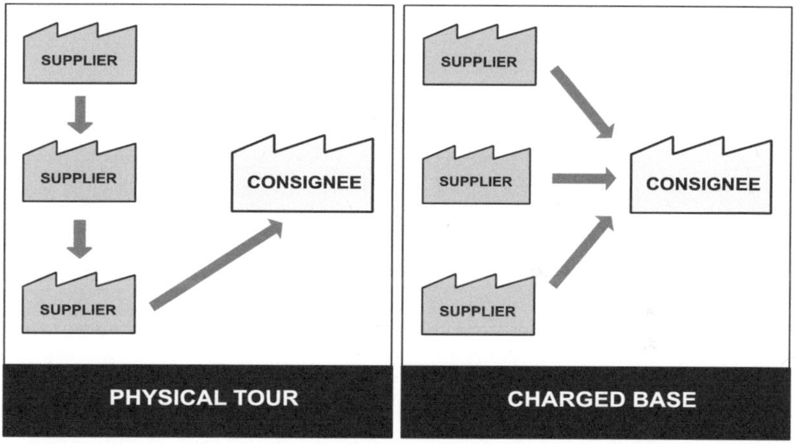

Figure 3.3: Physical tour and charged tour in area freight forwarding

Figure 3.4 depicts a swim lane diagram of the flow of information in case of an area freight forwarder concept. This is organized as follows. The consignee places an order, usually the result of an Material Requirements Planning (MRP) run, at the supplier (see section 3.2.1). The order contains information regarding which part number is required in which quantity at what arrival date. Based on the transport lead time, the supplier calculates a pickup date and then orders a transport by the carrier. The transport order consists of information regarding the shipment size (number of pallets, total weight) and the pickup date. Afterwards, the carrier picks the shipment up at the supplier and delivers it to the consignee.

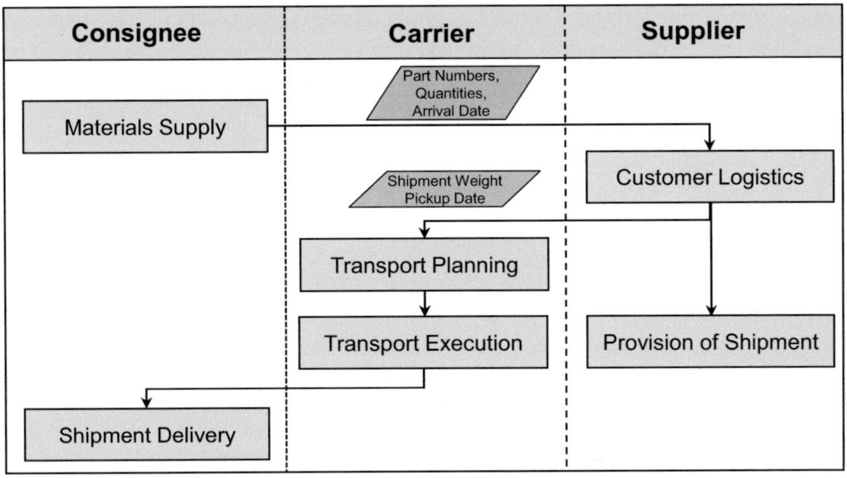

Figure 3.4: Communication between consignee, carrier and supplier in case of an area freight forwarder concept

One advantage of the concept is the low effort of planning and control of the transports for the consignee. These tasks are all outsourced to the carrier who takes care of all the different tasks. Not even the scheduling of the pickup needs to be performed by the consignee, since it is accomplished by the supplier. Another advantage of the concept is the aforementioned pooling effect. Since carriers which operate as area freight forwarder usually contract with companies from different sectors, their total demand for transport capacity is the sum of various independent random variables. Therefore, there is substantial pooling effect which results in a high utilization of the fleet.

In contrast to the advantages, the concept has some disadvantages. Due to the flow of information, the concept is intransparent to the consignee. This intransparency has a detrimental effect regarding the reliability of the concept and the service he pays for.

Between the placement of the order and the delivery of the shipment is usually a lead time of several days. During this lead time, the consignee gets no feedback or status regarding the delivery. Moreover, there is no communication between the goods in and the materials supply department of the receiving plant in order to check if the material arrives on the requested delivery date. If, for instance, the supplier did not receive the order, the consignee will only discover this when the material which has been ordered is not in stock. In this case, an extra tour, usually an expensive express delivery, is required to procure the material as quick as possible.

Another disadvantage is that the consignee does not know about the tours the carrier conducts and thus does not know for which service he pays for. This information asymmetry enables the carrier to take advantage of the consignee.[1] It can be illustrated by the following examples.

In a simple case, the consignee orders a shipment which is large enough to fill a whole truck. The carrier physically performs a direct transport, but does not charge according to a "direct transport" tariff. Instead he charges a higher price according to the AFF tariff, which is higher than the direct transport tariff. Since the consignee usually does not know about the size of his shipments, he is not able to detect this.

Another example is if several plants of the consignee place orders at the same supplier. The orders all pass the same hub of the carrier and can thus be regarded as one large shipment. Since the price increases degressively with the shipment size, there would be a discount. However, since the consignee plants do not communicate with each other, the carrier charges different small shipments, which results in a higher price.

[1] In microeconomics, this is called *hidden action.* The principal (the consignee) is not able to completely observe the task which is performed by the agent (the carrier). Since the agent can take advantage of the principal without getting detected, this is called *moral hazard.* (Varian 2014)

3.1.3 Milk Run

The third standard transport concept recommended by VDA 5010 are so called milk runs. As milk runs are a special case of groupage service transports, their goal is to consolidate LTL shipments from multiple suppliers to create bigger loads. Therefore, the delivery frequency can be increased without decreasing vehicle utilization (cf. Figure 3.5). The high delivery frequency or small lot size, as shown by the EOQ-Model[2], reduces cycle inventory. (Harris 1913)

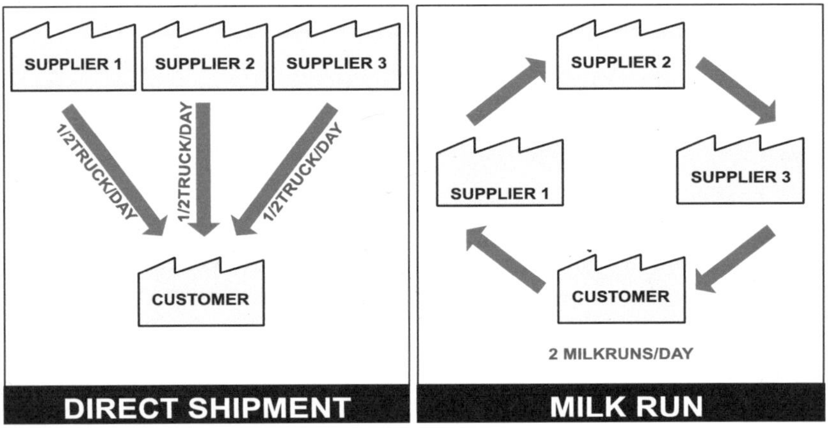

Figure 3.5: The idea of the supplier milk run

The term "milk run" derives from the traditional American and British milkman. The milkman had a fixed daily route on which he simultaneously delivered milk and collected the empty bottles. Sometimes, the empty bottles were even used as a trigger to leave some fresh bottles at the respective house.

2 The so-called Economic order quantity model finds on optimum order lot size taking into account the tradeoff between inventory and lot size.

According to VDA 5010, milk runs visit more than one supplier and then travel directly or with transshipment to the receiving plant. Meyer (2015), further extends this definition by ascribing the following characteristics to milk runs:

- Fixed routes with fixed pickup and delivery time slots, fixed volumes and fixed cycles
- LTL shipments of two or more suppliers are consolidated to bigger loads. This opens the possibility of increasing the delivery frequencies without decreasing vehicle utilization.
- Full and empty returnable containers are replaced in a ratio of 1:1
- Milk runs are planned by the consignee.

Moreover, Klug (2010) differentiates between static and dynamic milk runs. In the case of static milk runs, the cycles and routes are fixed and the volumes are constant. Dynamic milk runs, in contrast, have changing routes, changing pickup cycles and changing volumes. The difference between a dynamic milk run and the groupage service and dynamic milk runs are that the dynamic milk run is planned in advance. That is, on the first day, we already know which tour is going to be performed on the remaining days of the planning period. In case of a groupage service, the AFF plans a new tour every day. The tours of the subsequent days are not known, i.e. there is no planning in advance.

Since milk runs are transport concepts which provide some degree of regularity, it facilitates planning for all the involved partners. The suppliers, which provide material, usually know at which time the milk runs arrive, since the same tour is repeated every day. This reduces errors in the shipment process. Further, the suppliers are able to plan the capacities in shipping area according to the work load. The logistics service provider usually encounters less waiting times in the suppliers' shipping area. Waiting times are usually caused by missing materials or other errors in the process of material provision.

Since these are reduced, the LSP encounters less waiting time. The consignee profits from well utilized transports and transparency regarding the service he receives from and is being charged by the logistics service provider.

However, although being a standard described by the VDA, and despite of the described advantages, milk runs do not play a significant role in Europe for inbound freight (see (Queiser 2007)). The concepts of direct shipments and area freight forwarding are predominant. According to Meyer (2015), the reasons are both operational cost and operational complexity.

Selecting an appropriate capacity is a special challenge in planning milk runs for two reasons. First, in many companies, the master data are inaccurate. This means, important data to calculate the necessary capacity such as weight and volume are not known. This is why it is not possible to calculate the required capacity which is required for designing a milk run. If the master data is accurate, we can calculate the required capacity but it is rather complicated due to the variability of demand. Since demand is variable, the transport capacity which is needed is also variable. That is, there are days with a higher demand and days with a lower demand. If the demand for multiple parts are correlated, the effect is amplified. When we plan a milk run, we select a vehicle which has sufficient capacity to fulfill the transport orders with an acceptable service level. The higher the variability, the lower the utilization of the capacity (see the law of variability buffering in section 2.4). This is why milk runs are usually employed in case of parts which are subject to a low variability in demand (cf. section 3.3).

3.2 Control Policies of Materials Supply

The preceding section reviewed the physical flow of material which is usually involved in systems of transport logistics. In this section, the flow of information is reviewed. We give an overview of the fundamental concepts and models which are relevant in procurement logistics and material supply planning respectively. After a short introduction regarding the basic principles governing material flow, the first section of this chapter explains the concept of material requirements planning which is still widespread among the German automobile industry. Afterwards we present stochastic inventory policies, which are sometimes also applied in association with stochastic inventory policies. Afterwards, we explain the kanban system, which is control policy that originates in lean production systems.

There are two distinct principles, by which the flow of material can be controlled: push and pull. As the terms indicate, in a push strategy products are "pushed" into the system by an upstream process. In a pull system, they are "pulled" by the customer, i.e. a downstream process. Hopp and Spearman (2011) give a more exact definition: *"A push system schedules the release of work based on demand, while a pull system authorizes the release of work based on system status."*

Therefore, in a push system we need an exogenous schedule (e.g. a production plan) to release work into a system (e.g. a production system). The release time is rigid, i.e. it is not modified according to what is happening in the process itself. In a pull system, a job is released when a signal is generated by a change of the system state, e.g. a change in line status.

Since the terms "push" and "pull" are not defined uniformly in literature and the terms are partially utilized in a conflicting manner, Hopp and Spearman (2004) published a paper which further elaborates the discussion and give some important distinctions. One important criterion which distinguishes push and pull systems is that the amount of work in process (WIP) is always limited in pull systems. That means an MRP system which puts a limit on the WIP is a pull system. Moreover a, base-stock system is – surprisingly – not a pull system. The reason is that backorders can increase infinitely beyond the base stock level. That means, the WIP can also increase infinitely. If we consider a base-stock system with limited backorders, i.e. lost sales, it becomes a pull system.

3.2.1 Material Requirements Planning

The first control policy we present is material requirements planning (MRP). In Material Requirements Planning, the trigger of material ordering is expected future demands, i.e. a forecast, for finished products, or so called end items. Therefore, it belongs to the group of push systems.

These expected future demands are stored in a so called Master Production Schedule. It contains the quantities and due dates of all end items and external demand for lower-level parts. Each end items consists of some lower-level items. The relationship between lower-level items and end items is described by the bill of material (BOM). With the information from the bill of material, the demand for lower-level parts can be calculated. This is called BOM explosion. (Chase and Jacobs 2017))

After taking into account all the inventory which is on hand or in transit, we can calculate the net demand for lower-level items. To determine the time the order is placed at the supplier, a deterministic lead time is used.

The MRP calculation for every part can be summarized by the following steps (Hopp and Spearman 2011):

The first step is called netting: The Master Production Schedule contains the customer's gross requirements for end products. If we already have on-hand inventory of the requested items, we need to subtract them from the gross requirements. If we already placed an orders (purchase or manufacturing), which has not been fulfilled yet, we also need to subtract them from the gross requirements.

The netting calculation yields the net requirements per part. To form jobs, the demand is divided into lot sizes by a lot-sizing-rule, e.g. the Economic Order Quantity or the algorithm of Wagner and Whitin (1958).

The jobs now need to be scheduled. This is done by simply offsetting the due date by a certain deterministic planned lead time. The offset due date is called order release date.

To perform the job, usually several lower-level-items are required. The information about which lower level items are needed in which quantity is stored in the BOM. By BOM-explosion, the demand for lower-level items is calculated. The results yields the master production schedule (i.e. gross requirements and due dates) of the upstream process.

This procedure is repeated for every process in the supply chain. Moving further upstream, at some point, we no longer create manufacturing jobs but place purchase orders at the suppliers.

An example of the aforementioned procedure is depicted in Table 3.2. We have a given Master Production Schedule (gross requirements and due dates) for some generic Part A, which can be an end item or a lower-level item. Our lot size is assumed to be preset to 75, our planned lead time is one period (e.g. one week) and we start with an on-hand inventory of 30. For simplicity, we do not take into account scheduled receipts.

The netting procedure yields that the initial on-hand inventory is sufficient to cover the gross requirements of period one. However, in period two we have a projected shortage of 5 units. Therefore we schedule the receipt of an order of 75, i.e. our preset lot size, in period two. Subtracting the 5 units from 75 yields 70 units. This is sufficient to cover the gross requirements of periods three and four. The remainder in week five will be 10, which is insufficient to cover the required 30 units. Therefore we schedule a receipt of additional 75 units in period five. The remainder for period six will be 55. Subtracting 30 for period six and 30 for period seven will result in a shortage in period seven. Therefore, we schedule another arrival of 75 units to period 7.

Table 3.2: Material Requirements Planning (c.f. Hopp and Spearman 2011)

Part A	Period								
	0	1	2	3	4	5	6	7	8
Gross Requirements		15	20	50	10	30	30	30	30
Projected on-hand inventory	30	15	-5						
Net requirements		0	5	50	10	30	30	30	30
Planned order receipts			75			75		75	
Planned order releases		75			75		75		

As described above, after netting and lot sizing, we have to schedule the order release dates. Our planned lead time (e.g. for a purchase or manufacturing order) is one period. Therefore, each order release date must be one period before the due date.

As intuitive and simple as it may look in theory, Hopp and Spearman (2011) state several shortcomings of the MRP procedure which lead to severe problems in practice. Firstly, MRP assumes deterministic lead times for production lines. However, queuing theory and Little's Law (cf. Little 1961) tell us that this assumption is invalid. Lead time is not constant but rather depends on the utilization of the system. Keeping the possible throughput (i.e. capacity) constant and increasing the system load results in higher lead times. Vice versa, decreasing the system load decreases the lead time.

Since planners are forced to assume a deterministic planned lead time, which is actually stochastic, they usually choose a pessimistic estimate which is longer than the actual lead time. That means, jobs are scheduled systematically too early, resulting in excessively high inventories.

Another aspect of criticism is that the MRP system is very sensitive regarding changes in the master production schedule, i.e. the planned order releases are subject to large changes. That is, the plans which are made by the system are actually quite good. But in practice, it happens that these plans fail (i.e. due to machine breakdowns), resulting in further plan changes and poor performance of the production system.

The orders which are generated as a result of the MRP run are usually sent to the suppliers as so-called call-off orders in regular time intervals (e.g. daily or weekly on the same day). In the German automobile Industry, suppliers are usually contracted by means of a framework contract, so-called delivery schedules ("Lieferplan"). In this framework contract, the parties agree on a total quantity of parts which need to be delivered in a certain period of time. The total volume of this agreement is that high that the order cannot be fulfilled on a single day but is split in smaller orders over the total duration of the framework agreement. The exact quantities which are needed each day are then "called off" by these call-off orders.

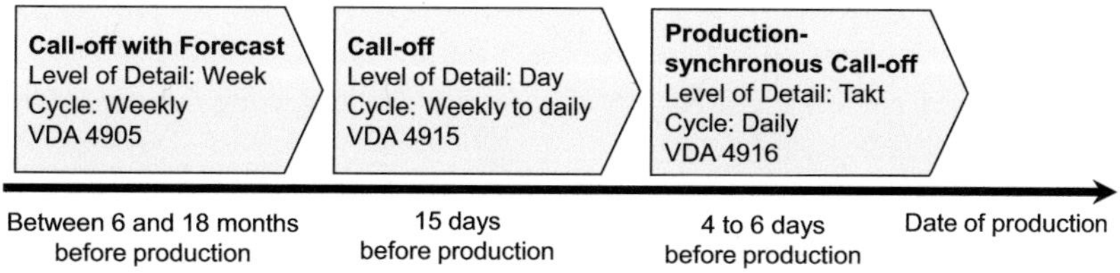

Figure 3.6: Call-off systems in the German Automobile Industry (cf. Klug, 2010)

VDA specifies three different standards for these call-off orders. The simplest and most widespread form is specified in VDA 4905, the call-off with preview. Each call-off order consists of the order for the next period and a preview of the following periods. This preview is for a minimum period of 6 months up to 18 months. The quantity, which is relevant for the fulfillment of the order by the supplier, is the order quantity for the direct subsequent period.

VDA 4915 specifies the short term call-off order ("Feinabruf"). It contains further information regarding the requirement for each part, e.g. the exact time of day the part is needed. Because of the high level of detail, it only contains a preview of two weeks. It is transmitted at least once week and at most once per day. In case of just-in-sequence delivery, the production synchronous call-off is employed. It is updated multiple times a day and contains the exact demand for parts based on actual production orders and information regarding the sequence.

3.2.2 Stochastic Inventory Policies

Another group of control policies employed in the German automobile industry are stochastic inventory policies. According to Tempelmeier (2011), an inventory policy consists of a set of decision rules, which determine the future orders. That is, the policy determines which quantity has to be ordered and when the respective quantity needs to be ordered.

Inventory policies are defined by at least two of the following parameters:

- The review interval (or order cycle) r defines the time between two inventory reviews. r=2 for instance stands for reviewing the inventory every second period.
- The order-up-to level S ("Big Stock") stands for a target inventory level. Each time we review the inventory and place an order which increases the inventory position[3] (physical inventory + outstanding orders – backlog) to the order-up-to level.
- The installation stock s ("small stock") is an inventory level which, upon reaching, triggers the immediate placement of an order.

Table 3.3 distinguishes the different types of inventory policies by combining these parameters. The time between orders can be either constant (e.g. in the case of a review period) or variable (as in the case of an installation stock). The order lot size can also be constant or variable.

[3] In inventory management literature, the term inventory position is sometimes called "disposable inventory".

Table 3.3: Overview of stochastic inventory policies

		Time between Orders	
		Constant	Variable
Order Lot Size	Constant	(r, q)	(s, q) (r, s, q)
	Variable	(r, S)	(s, S)

The first parameter combination in Table 3.3 is constant order lot size and constant review interval, i.e. the (r, q) policy. According to this policy, we order a quantity of q units every r time units. We always order q units regardless if there was a prior consumption. Therefore, this policy is not consumption-driven.

In the case of a constant order lot size but variable time between orders, we can choose between the (s, q) and the (r, s, q) policy. In the (s, q) policy, the inventory is reviewed continuously. Each time the level of physical inventory falls below the reorder point, we place an order of q. One alternative – which is far more widespread in practice – is the (r, s, q) policy. In this case, we review the inventory every r time units and order if the inventory position is below the reorder point.

In case of variable order lot sizes, there are the (r, S) and the (s, S) policy. Both policies are order-up-to S policies. However, in case of the (r, S)[4] policy we review the inventory position every r time units and then place an order which increases the inventory position up to s. In case of the (s, S) policy, we continuously review the inventory position. Each time it falls below the reorder point, we place an order which increases the inventory position up to S).

[4] Note that the (r, S) policy is closely related to the so-called base stock policy, the latter being a special case of the former (r=0, i.e. continuous review) (Silver et al. 1998)

Note that, as discussed in beginning of the section, the (r, S) and the (s, S) policy with unlimited backorders are no pull-based policies according to the definition of Hopp and Spearman (2004): If we order an infinite amount of material from the inventory system, this also results in an infinite amount of backorders and thus an infinite amount of WIP in the upstream process. Therefore, it cannot be a pull-policy. However, we assume the amount of backorders in these policies to be limited (i.e. lost sales occurring). This also leads to a limitation of WIP in the upstream process. Therefore, the (r, S) and the (s, S) policies with limited backorders are pull-policies.

Figure 3.7 depicts the mechanisms of the (s, q) policy in a graph of inventory over time. The continuous line marks the physical inventory over time, the dashed line marks inventory position over time.

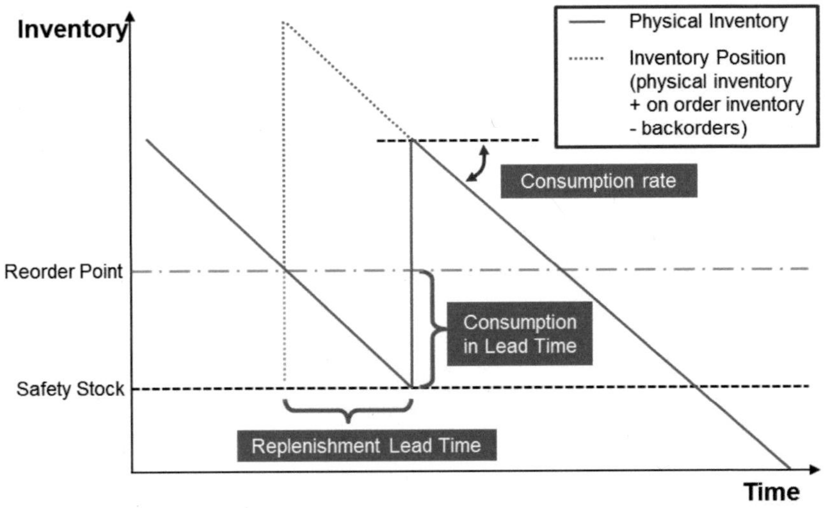

Figure 3.7: Illustration of the (s, q) policy (cf. Tempelmeier 2011)

Once the physical inventory falls below the reorder point, a replenishment order is triggered. In that instant, the open orders and hence the inventory position increases by the order size q. During the replenishment lead time, the level of physical inventory further decreases and approaches the safety stock. When the replenishment order arrives, the open orders are instantly reduced by the amount that has arrived. Hence, the physical inventory and inventory position become one identical line again. The goal is to set the reorder point that high that inventory does not fall below zero for a desired level of statistical safety.

3.2.3 Kanban Systems

Another implementation of a pull system is the so-called kanban system. It was originally developed at Toyota and is a famous element of the Toyota Production System. The word “kanban” is Japanese and means “card” in English. It stands for the signal which triggers the flow of material in the production system. (Hopp and Spearman 2011)

The idea of the kanban system is to organize the flow of material in a production system just like the process of shelf-stocking in a supermarket (e.g. a grocery store). That is, the customer takes the product he likes from a shelf. The taken product leaves a gap on this shelf. When the gap is noticed by an employee, the shelf is re-stocked. (Ohno 1988)

This concept has been transferred to production systems. The term “supermarket” has not been altered. In a production context, the term is used for an inventory store close to production lines. They are employed as buffers whenever a continuous flow (or one piece flow) is not possible. Moreover, the definition of the term customer has been extended. In a production system, the term customer does not only stand for the final customer who receives the final product. Therefore, in a process chain, each process is regarded as the customer of its upstream process and the supplier of its downstream process. (Rother and Shook 2009)

The mechanism of kanban systems is depicted in Figure 3.8. Each process is adjacent to an upstream and a downstream supermarket. Each time the customer consumes a units from the downstream supermarket, the process receives a signal (i.e. a kanban card) to replenish the consumed product. In order to replenish the product, the process consumes material from upstream supermarket, triggering a signal for replenishment to the upstream process. (Hopp and Spearman 2011)

As described before in this section, in a pull system the flow of material depends on the system state, e.g. the stock level in a certain echelon. If one unit is taken from this stock, a kanban which triggers the replenishment of the consumed unit is sent to the upstream process.

The benefit of the system is that processes only replenish material if there has been consumption before. Without prior consumption of material, no kanban cards are sent back to the upstream process. Therefore, the work in process (WIP) inventory is limited to the number of kanban cards. As a consequence, the lead time of the systems becomes more predictable. (Hopp and Spearman 2011)

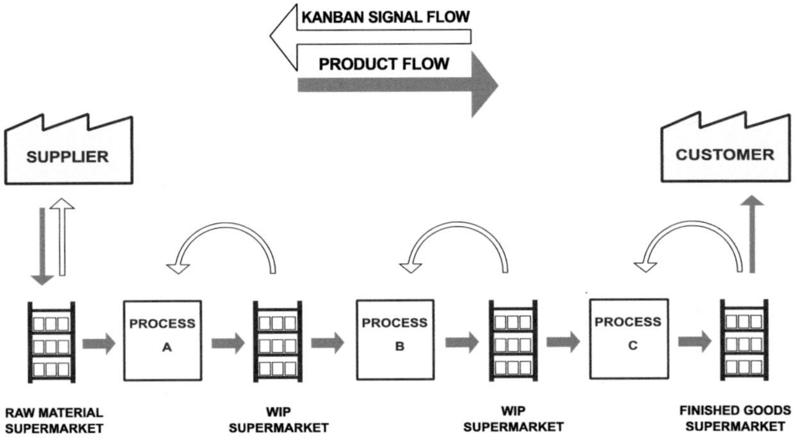

Figure 3.8: Mechanism of a kanban system (cf. Hopp and Spearman 2011)

Initially, kanbans were only used to control the flow of information in production systems, i.e. inside the plants. Eventually, with increasing maturity, the concept was extended to control the ordering of raw materials at the suppliers. (Baudin 2005)

In contrast to the case of manufacturing, in which kanbans are physically transmitted/transported over the shop floor, the information is usually transmitted electronically in the case of supplier kanbans. That is, internally, the cards are collected in a mailbox and in regular time intervals the information regarding the consumption is sent to the supplier via email or fax. (Baudin 2005). Another way would be to completely virtualize the kanban cycle by applying the concept of electronic kanbans (e-kanban)[5].

The mechanisms of a supplier kanban system are depicted in Figure 3.9. At the receiving plant, there is a supermarket which is filled with material that needs to be ordered from an external or internal supplier. When the customer, in this case a production process, withdraws material from the supermarket, a replenishment is signal in form of a kanban is triggered and sent to the supplier.

It is both possible that the signal is sent instantaneously in real time to the supplier, or that the number of free kanbans is counted at regular time intervals and the information is sent to the supplier electronically. The supplier provides the material which is then picked up by a logistics service provider and transports the material (see section 3.1) to the receiving plant.

[5] In case of e-kanban, the physical cards in a kanban cycle are replaced by virtual inventories and orders while limiting the WIP inventory to a certain amount.

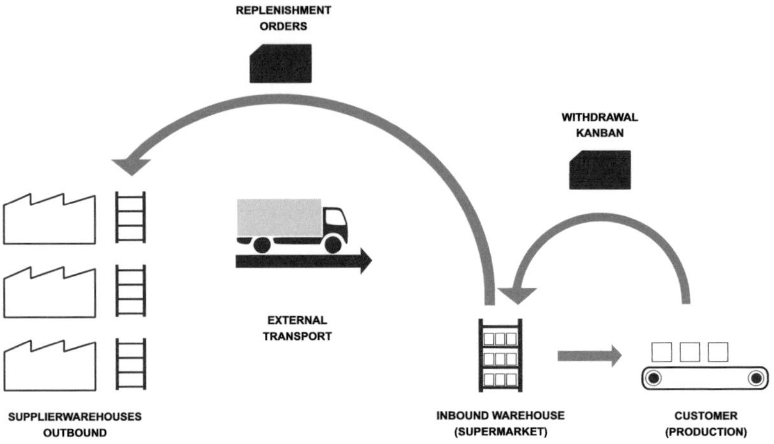

Figure 3.9: A supplier kanban system (cf. Furmans 2007)

3.3 Combining Transport Concepts and Control Policies

The preceding sections presented the physical flow of material and the associated flow of information which are employed in systems of transport logistics. In this section, we investigate the interaction between the two subsystems and their synthesis, i.e. how they can be combined. Moreover, the need for measurements of stabilization in transport logistics is pointed out.

VDA 5010 gives some qualitative advice regarding the choice of an appropriate transport concept for a certain supplier, which is depicted in Figure 3.10. To facilitate the decision, various decision criterions are given:

Regularity of Transports The variability of the time between two orders that are places at the supplier. A high delivery frequency with constant interarrival times corresponds to a high regularity. A low delivery frequency with varying interarrival times corresponds to a low regularity.

Load Structure The size or capacity of the load. If the load is sufficient to fill up a whole truck, we are dealing with a full truck load. In case of a smaller load, we speak of a less then truck load shipment.

Distance to other Suppliers In case of other suppliers being in the proximity of the supplier whose transport concept is to be assigned, the distance is short. That is, we can combine the adjacent suppliers to a route.

Stability of Required Transport Capacity The variability or the probability of the required capacity being below a certain bound (cf. section 2.5).

Easy Load Consolidation The ability to consolidate the load from less than truck load to full truck loads.

The figure shows that for all full truck load shipments, regardless whether the shipments are regular or not, the direct shipment is the appropriate transport concept. In case of less than truck load shipments, the area freight forwarder is the concept of choice. Only in case of high load stability, short distance to adjacent suppliers and easily possible load consolidation, the advice is to employ milk runs for procurement of the parts.

The predominant concept for LTL shipments in the German automobile industry is the area freight forwarder concept (Meyer 2015, Schöneberg 2010). The reason for this becomes obvious by taking into account the advice given in Figure 3.10 and the fact that many receiving plants still plan their production following the MRP planning concept (cf. section 3.2.1).

The MRP planning concept, however, is susceptible to the bullwhip effect, i.e. minor fluctuations of customer demand amplify moving upstream the supply chain (see section 2.2.1). That is, the replenishment orders which are placed at the suppliers are in the best case just as variable as the customer demand, but usually much more variable. That is, the level of stability of the replenishment orders is low (cf. Figure 3.10). (Alicke 2005)

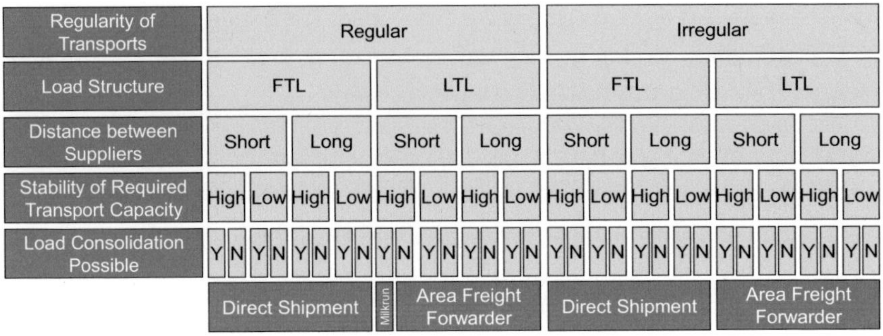

Figure 3.10: Decision tree for choosing the appropriate transport concept (VDA 5010)

Even if the receiving plant levels its production according to the heijunka philosophy (cf. Veit 2010 and section 4.1), the variability can usually not be totally eliminated. The so-called pacemaker process is usually far downstream. In between, there are several processes steps before we get to the interface to the supplier. Directly before the pacemaker, the demand is leveled. But moving further upstream, lot sizing restrictions due to setup times, fluctuations of machine availability and bills of material re-inflate the variability of demand, causing a bullwhip effect inside the plant.

As stated by Meyer (2015), both the milk run concept (cf. section 3.1.3) and the kanban concept are concepts that originate in lean manufacturing, i.e. the Toyota Production System and should ideally be applied together. However, we identify certain limitations of the plain kanban control logic which makes it – in many cases – unsuitable for a combination with milk runs.

A milk run tour typically serves two or more suppliers, each one delivering one or more part numbers. The demand for each part number is stochastic and fluctuates with a higher or lower variability. Each part also has a requirement for transport capacity (i.e. a certain weight or volume). The total capacity for each tour on each day is the sum of the capacities requirements of all part numbers.

If the variability of the demand for these parts is low, it is possible to achieve a high vehicle utilization. Since the dispersion of the capacity distribution is low, the vehicle is often well-utilized and only seldom poor-utilized. If however, the variability of demand increases, we need to increase the vehicle capacity in order to buffer the variability (cf. section 2.4). The result is a lower vehicle utilization. Therefore, we conclude that the economic efficiency of combining a kanban system with milk runs heavily depends on the variability of the demand for parts. This effect is depicted in Figure 3.11.

As we pointed out before, many firms in the German automobile industry still plan their production schedule with MRP. That is, the shipment capacity requirement is only rarely stable and milk runs would be uneconomic. Therefore, a large portion of their transports is operated according to the AFF concept. (cf. Figure 3.10)

As described in section 3.1.2, the AFF concept has some advantages for the consignee.

- All the planning tasks are **outsourced to a specialist**, which is lenient for the consignee.
- The AFF delivers material to different customers from different industries. The **pooling effect** (cf. section 3.1.2) levels the total transport capacity which needs to be provided by the freight forwarder
- AFF are organized in coalitions. In case of **peak demands**, they can subcontract peer companies
- In case of **slump demands**, they can simply skip tours and leave vehicles at the depot.

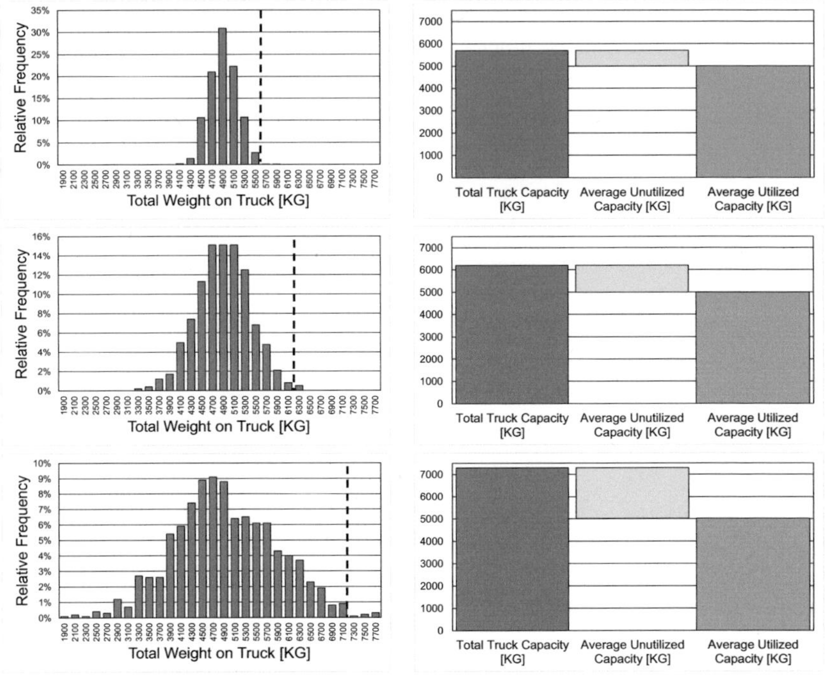

Figure 3.11: Capacity requirement of low and high variability parts (μ=5000, cv^2= {0.05, 0.1, 0.2})

However, the concept has some limitations, which are detrimental for the consignee and the environment. First, the lack of transparency regarding the billing of services enables the freight forwarder to exploit the consignee. The freight forwarder charges all loads according to his AFF tariff table. It is however possible, that a load is that big that it fills a whole truck. Instead of charging the lower price for a LTL direct shipment, he is able to charge a large weight in the AFF tariff. (cf. section 3.1.2)

Another limitation is the fact that since the transport capacity required from the freight forwarder is subject to variations, the road infrastructure is also variably utilized. In case of peak demand, a high amount of transport capacity is needed. That means, many vehicles are using the transport network, increasing the possibility of congestion. In case of slump demand, only few vehicles are utilizing the transport network, i.e. we observe under-utilization.

In case of a milk run or direct transport concept, these disadvantages do not occur. In case of a milk run, the freight is not charged to the actual weight or load volume. The consignee pays for the whole truck which performs a defined tour at defined times. Therefore, the service is very transparent. Moreover, since transports are performed regularly and there is no ad-hoc adjustment of transport capacity to cover peaks, an increasing dissemination of milk runs reduces the utilization of the road network. This has many desirable for society, i.e. congestions. In order for the milk run to be efficient, we need stability regarding the required transport capacity.

This is why, in the following we want to elaborate a concept for the stabilization of the demand for transport capacity. In section 2.4 we stated that in order to achieve stability, we can directly decrease process-inherent stabilization or employ buffers. The buffer must be either capacity, inventory or time.

The demand for transport capacity is directly linked to the customers demand for end products, which is stochastic in nature. This is why we are not able to decrease process inherent variability and to achieve stability, we need to employ buffers.

As mentioned before in this section, the demand for transport capacity is variable and there is no damping mechanism. The variations are handled by holding enough vehicle capacity available. Thus, we currently buffer the variability by reserve capacity, which is undesirable.

In our view, it is more desirable to shift variability from the capacity dimension into the inventory or time dimensions. In production logistics, this is facilitated by heijunka leveling. Therefore, in the following we will show how heijunka leveling can be employed in transport logistics to level the demand for transport capacity.

4 Leveling as a Concept of Design for Stability

Milan or Madrid - as long as it's Italy!
(Andreas Moeller, German football player)

Based on the description of transport logistics systems and their need for stability which were presented in the preceding section, this section elaborates a concept of design for stability in transport logistics. The section starts with a short description of heijunka leveling in production logistics. On this basis, we describe how the system needs to be adapted for an application in materials supply or transport logistics, respectively. Afterwards, the systems are modeled mathematically to understand the system behavior. The last subsection explains how these mathematical models can be applied to design the system.

4.1 Heijunka in Production Logistics

According to Furmans and Veit (2013), heijunka leveling is "a simple method for lot-sizing and production scheduling" using kanban cards. It is used to manage the production of a variety of products which share a common resource, i.e. a production line. In production logistics it is employed to control the replenishment of a finished goods inventory. In a heijunka board, production capacity is reserved for each variant which is produced on the resource, the so-called pacemaker process. By buffering, short-term fluctuations of customer demand are filtered out and the required production capacity is leveled.

The system mechanics are displayed in Figure 4.1.The customer places orders at a plant. The plant fulfills these orders by withdrawing material from a finished goods inventory. For each unit that is withdrawn from the finished goods inventory, a replenishment order ("production kanban") is released. The material is not replenished immediately since production follows a certain production sequence. In this sequence, a certain time slot is reserved for each quantity. Only in this time slot, the respective product is produced. Raw material, which is needed for the production of finished goods, is taken from a supermarket at the final assembly. All upstream processes are connected via kanban loops. (cf. Smalley et al. 2004)

The time slot in in the production sequence is limited so that we can only produce a certain maximum quantity. If the order of the customer exceeds the maximum quantity, the remaining kanban cards are put in an order buffer ("overflow"). When, at some later point in time, the customer orders a quantity which is smaller than the maximum quantity, orders from the order buffer are released.

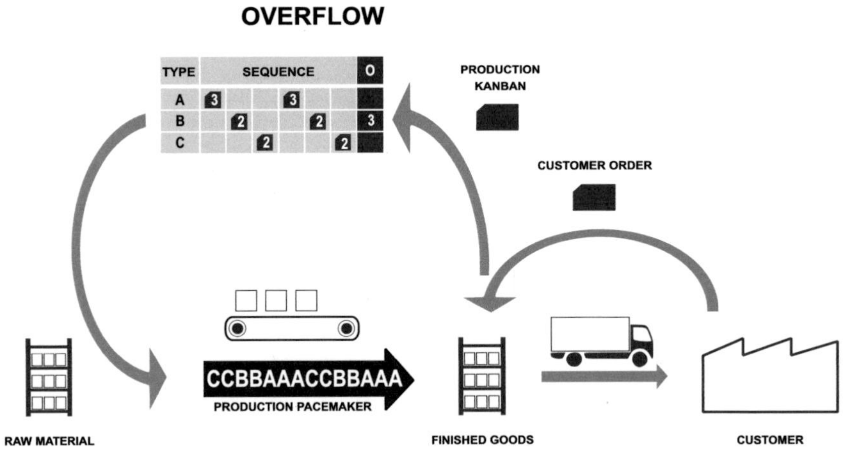

Figure 4.1: Heijunka leveling in production logistics (Furmans 2007)

The interaction of the finished goods buffer and the order buffer result in the desired effect of leveling. In case of peak demand, we simply withdraw more material, using the inventory buffer. In case of slump demand, we produce orders from the order buffer.

The effect is that the system produces the same quantity almost every day, which is beneficial for the upstream processes: Since the sequence defines a maximum production quantity for each product, the upstream process can rely on that this maximum quantity is not exceeded. Therefore, the supermarkets which connect the upstream processes need to buffer less variation because of the leveled consumption.

Table 4.1 displays an example of a heijunka sequence in a production environment.

Table 4.1: Heijunka pattern for three days in a production environment (EPEI=3d)

Part Data		Day 1		Day 2		Day 3	
ID	CT [min/unit]	Quantity [units]	Total Time [min]	Quantity [units]	Total Time [min]	Quantity [units]	Total Time [min]
A	5	105	525				
B	3	100	300				
C	1	60	60				
D	3			80	240		
E	4			150	600		
F	2			100	200		
G	1			130	130		
H	5					50	250
I	5	50	250			40	200
J	4					85	340
K	3					80	240
L	2					70	140
Required Time [min/day]			1135		1170		1170
Available Time [min/day]			1200		1200		1200

In the example, different parts are produced on a resource (e.g. a machine or production line) in a certain sequence. The Every Part Every Interval (EPEI), i.e. the duration of the sequence, is three days. After these days, the pattern recurs. Each part has certain requirements regarding capacity, which can be calculated by multiplying the cycle time per unit (CT) and the number of units which need to be produced. In order to be able to fulfill the customer's requirements, the required capacity must be smaller than the available capacity.

4.2 Heijunka in Materials Supply

In the preceding section we gave a brief summary of how heijunka leveling is used for the stabilization of production quantities. On this basis we now describe how the concept of heijunka leveling needs to be adapted to be transferred from production logistics to transport logistics.

Whilst heijunka is already described and industrially employed in production logistics, the concept is new to transport logistics. In section 4.1 our focus was on the interface between a producing plant and its customer. In this section, we move the focus upstream the supply chain, where we investigate the interface between a receiving plant and its supplier.

Figure 4.2 depicts the process usually employed in materials supply. The two entities of our system are the receiving plant, which orders and later receives the material, and the supplier, who sends the material. The interface of these entities is transport logistics. As stated in section 3.1, the transport can be organized according to different concepts. These are direct transport, area freight forwarding or milk runs. All of these concepts can be applied with heijunka leveling.

In our materials supply system, the receiving plant produces goods for its customer. To build these goods, raw material is consumed. After the consumption of raw material, an order policy triggers the replenishment of the goods which have been consumed. The supplier provides the material as ordered. Afterwards, the LSP picks up the material and delivers it to the receiving plant.

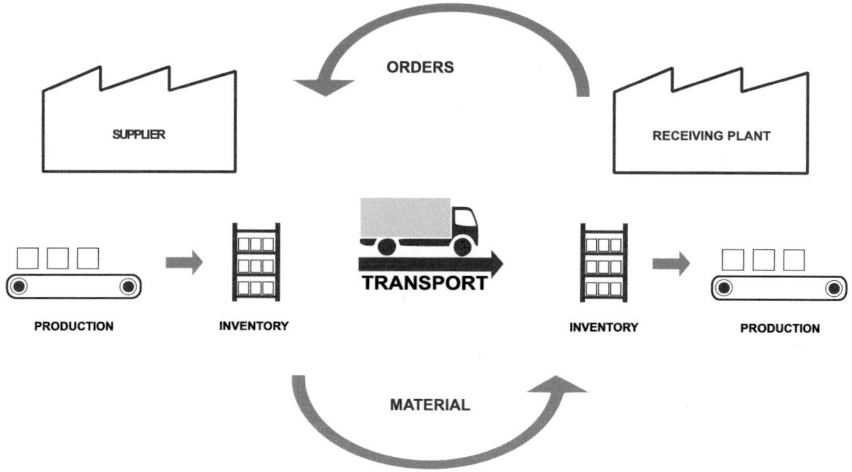

Figure 4.2: Logistics system in procurement

The goal of heijunka leveling in transport logistics is to stabilize the required transport capacity, i.e. reduce its variation. The capacity is a linear function of the capacity per unit and the respective quantity. Since the capacity per unit, e.g. weight or volume, cannot be varied, we need to reduce the variation of order quantities. The orders are controlled by a control policy, as described in section 3.2. Therefore, in order to stabilize the required transport capacity, we need a stabilizing control policy, which is provided by heijunka leveling.

Figure 4.3 depicts a system of heijunka leveled materials supply with a milk run who picks up material at three suppliers. Again, the production of the receiving plant consumes raw material which is stored in a buffer inventory or warehouse. After the consumption, the raw material needs to be replenished. For the replenishment, we place orders at the supplier.

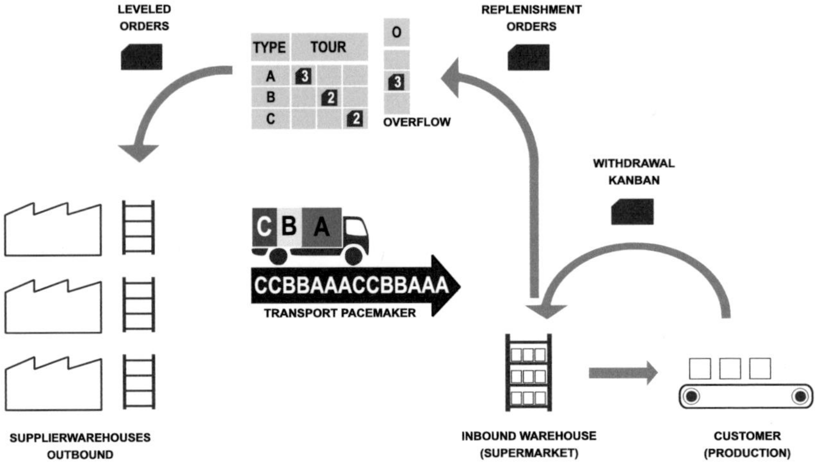

Figure 4.3: Heijunka leveled kanban system in materials supply

Similar to heijunka leveling in production logistics, the kanban cards withdrawn from the supermarket do not directly become replenishment orders but are first put into the heijunka board. In the heijunka board, there is a certain amount of capacity reserved for each part number (A, B, and C in Figure 4.3) on each day. That means, there is a limitation regarding the replenishment order size which can vary from day.

If the consumption of the production process exceeds this maximum order size, we can only order the maximum order size. The amount by which the maximum order size is exceeded is stored in an order buffer, the so-called overflow. If the order size is lower than the maximum order size, we may increase the replenishment order size up to the maximum order size, if there are enough orders in the order buffer.

This mechanism levels the replenishment orders. In the case of an upward fluctuation of raw material consumption, we put the excess quantity in our order buffer. In the case of a downward fluctuation, we add orders from the order buffer to our replenishment order so that we do not lose capacity. In effect, we order the same amount most of the time: the amount which corresponds to the capacity we reserved on our truck.

Since the truck capacity is constant for the duration of the leveling period and must not be exceeded, the reserved capacity slots for all the products need to be aligned with each other. Each day, the reserved capacity must not exceed the truck capacity.

If there is only one part number which is ordered on a regular basis and delivered by one truck, we have a special case of heijunka leveling: The whole truck capacity is reserved for one kind of material. This is the case if we order one part number from one supplier and the capacity requirements (quantity*capacity/quantity) are that high that we can utilize a full truck to procure the material in an acceptable order frequency.

In the case of multiple part numbers which are assigned to a truck, the different kinds of parts share the capacity of the truck. Thus, we a reserve a capacity slot for each product. It is possible that the different parts are picked up at multiple suppliers (milk run case) or only one supplier (direct transport).

In the case of area freight forwarders, the vehicle utilized to pick up goods at the suppliers might change from day to day. In this case, we can still place leveled orders at the suppliers and the freight forwarder can plan its capacity more easily.

Table 4.2 displays an example of how the heijunka pattern can look like in case of heijunka leveled material supply. Similar to production logistics, capacity in transport logistics is a linear function of the quantity. The difference is the proportionality factor. Whereas in production logistics, the capacity depends on the cycle time per unit, in transport logistics the capacity depends on the unit weight or unit volume. In the example, the payload of the truck performing the transport is 10,000 KG. Therefore, each day, reserved capacity must not exceed this maximum capacity. Again, the EPEI is three days and the pattern is recurring.

Table 4.2: Heijunka pattern in the case of leveled materials supply

Part Data		Day 1		Day 2		Day 3	
ID	Weight [KG/unit]	Quantity [units]	Total Weight [KG]	Quantity [units]	Total Weight [KG]	Quantity [units]	Total Weight [KG]
A	50	100	5000				
B	25	80	2000				
C	10	50	500				
D	30			200	6000		
E	40			50	2000		
F	20			50	1000		
G	10			75	750		
H	5					40	200
I	50	40	2000			40	2000
J	20					80	1600
K	30					100	3000
L	20					140	2800
Total Weight per Day [KG]			9500		9750		9600
Vehicle Capacity per Day [KG]			10000		10000		10000

A heijunka leveled transport logistics system works just like a milk run or direct transport works today. The consignee calculates the capacity he needs over a certain period of time for a certain tour and contracts the freight forwarder, who performs the tour as desired. The supplier is informed regarding the arrival times of the supplier and provides the material ordered by the leveled replenishment policy. Since we have reserved capacities for the whole leveling period, the plan can even serve as pickup sheet for the freight forwarder to create more robustness by checks at the shipment area of the supplier.

4.3 System Modeling

Based on the system description which was presented in the preceding section, we build mathematical models of the system in this section. The purpose of the models is to understand the system behavior and to enable us the design the system parameters. Therefore, we first show how the models of Veit (2010) and Lippolt and Furmans (2008) can be applied to model the inventory behavior in case leveled replenishment but variable consumption in an environment of transport logistics. Afterwards, we build an optimization model, which enables us to calculate a leveling pattern.

4.3.1 Inventory Behavior

To build a mathematical model of the system, we isolate one part number from the inbound supermarket of the receiving plant, which is depicted in Figure 4.3. This supermarket might contain various different part numbers. In the following, we only consider one of them.

Following Veit (2010) and Lippolt and Furmans (2008), we can describe the current state of this inventory by its deficit Z to a target inventory. In a kanban system, this target inventory is equal to the maximum inventory, i.e. the number of kanban cards in the cycle. The deficit is equal to the number of free kanban cards, i.e. the cards previously attached to parts and are currently waiting in the heijunka board or in transit. The current inventory and hence also the current deficit change over time. Consumption of parts decreases the level of inventory, replenishment orders increase the level of inventory (see Figure 4.4). Therefore, all state changes are results of the superposition of the stochastic processes of replenishment and consumption.

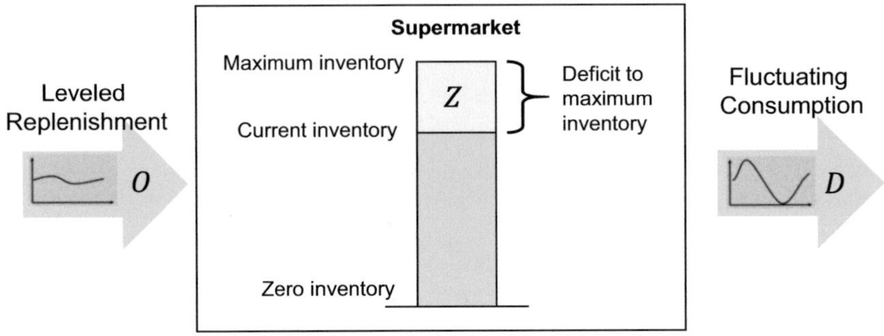

Figure 4.4: Description of the system state by the deficit to the maximum inventory

The deficit Z is the result of the superposition of the two stochastic processes of replenishment and consumption. Both the replenishment orders O and the demand D are independent, identically distributed random variables. In case of an order lead time of one period the deficit of the next period Z_{t+1}can be calculated by adding the orders of the current period O_t and subtracting the demand D_t from the deficit of the current period Z_t. In case of an order lead time, the orders arrive with a delay which we take into account by an offset O_{t-l+1}. The minimal value of Z is 0, that is, the current inventory is equal to the maximum inventory.

$$Z_{t+1} = max\{0, Z_t + O_{t-l+1} - D_t\} \tag{4.1}$$

The replenishment orders depend on the number of free kanban cards and the available capacity in a certain period. If the number of free kanban cards, i.e. the deficit, is greater than the reserved truck capacity, we only place a replenishment order at the supplier which is equal to the maximum allowed order quantity.

$$O_t = min\{Z_t, C_t\} \tag{4.2}$$

Given the deficit Z_t and the number of kanban cards N_K, we can calculate the current inventory I_t. In a kanban system, the maximum deficit is equal to the number of kanban cards N_K. Moreover, the minimum inventory is zero. If the inventory falls below zero, the system has a backlog B_t. Mathematically, this is expressed as follows:

$$I_t = max\{0, N_K - Z_t\} \tag{4.3}$$

$$B_t = max\{0, Z_t - N_K\} \tag{4.4}$$

With these equations, we can simulate the system, e.g. with common spreadsheet calculation software, or calculate the probability distribution of the inventory. Moreover, we can calculate important performance measures of logistics systems such as service levels.

A simulation is not always efficient which is why we might want to determine these values analytically. For the analytical calculation, we use an analogy between inventory systems and queuing system as proposed by Güllü (1998). Furmans and Lippolt (2008), Veit (2010) and Matzka et al. (2012) transfer this analogy to discrete time and use a G|G|1 queue to describe the behavior of the heijunka leveled kanban system. According to this analogy, the deficit (before demand) behaves a like the number of waiting customers in a discrete time G|G|1 queue.

Figure 4.5 depicts the relation between deficit and inventory for a numerical example with $N_K = 6$ and $Z = 2$. Since the number of kanbans is six, the maximum inventory is also 6 units. If the deficit to the maximum inventory, i.e. the number of customers in the queue, is 2, then the current inventory I must be 4. If a customer has been served, the number of units in the queue decreases by 1. This means that we have one more unit of physical inventory. If a customer arrives at the queue, the deficit increases by one unit. This means that the inventory has decreased by one unit. If the queue is empty, the deficit to the maximum inventory is zero, thus the current inventory level is equal to the maximum inventory N_K. If the number of customers in the queue exceeds N_K, the system encounters a backlog.

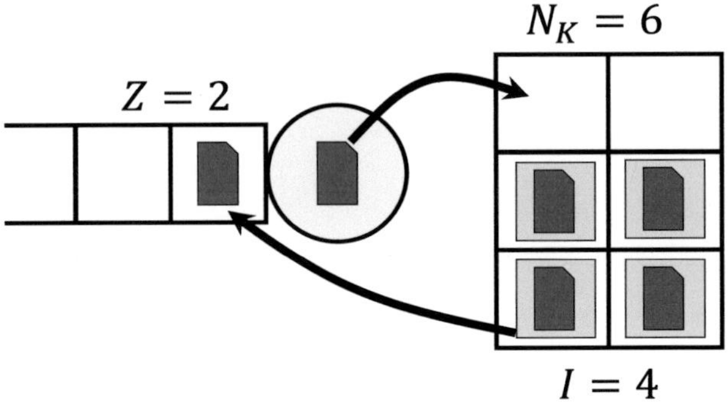

Figure 4.5: Relation between queuing system and inventory

The interarrival time distribution of the queuing system corresponds to the capacity distribution of the inventory system, i.e. the probability distribution of the number of units replenished within one discrete time step. In addition, the service time distribution corresponds to the demand distribution of the system i.e. the probability distribution of the number of units consumed within one discrete time step. If we increase the capacity of the system while keeping the demand constant, the inventory increases, hence the deficit decreases. In a queuing system, this corresponds to an increase of interarrival time while keeping the service time constant. This results in an increase in queue length, which, as stated above, corresponds to the deficit.

This analogy is helpful to understand the system behavior, since we can use it to calculate the probability distribution of the deficit for given demand and capacity distributions (see Figure 4.6). Equation (4.1) is equivalent to Lindley's equation in discrete time (Arnold and Furmans 2009). Therefore, an efficient method to quickly calculate the probability distribution of the deficit (which corresponds to the customer waiting time distribution of the queuing system) is provided by the first algorithm of Grassman and Jain (1989).

The first step is the calculation of the work balance u_j. This is done by subtracting the random variables of demand $\vec{d}=(d_0, ..., d_g)^T$ and capacity

$\vec{c}=(c_0, ..., c_h)^T$. For both vectors, we employ the discrete time notation, i.e. $P(X = i) = x_i$ for any discrete random variable X.

$$\vec{u} = \begin{pmatrix} u_{-h} \\ \vdots \\ u_0 \\ \vdots \\ u_g \end{pmatrix} = \vec{d} * rev(\vec{c}) \tag{4.5}$$

In equation (4.5), $rev(\vec{c})$ denotes the reverse of the capacity vector $\vec{c}$, i.e. we swap the sequence of the vector elements from $c_{min} \rightarrow c_{max}$ to $c_{max} \rightarrow c_{min}$. The work balance can be split in ascending $(u_1, \dots, u_h)$ and weakly descending $(u_{-g}, \dots, u_0)$ ladder heights. All elements of u for $j < -h$ and $j > g$ are zero.

We perform the iterative algorithm with the following steps:

- Initialize β_j^0=0 for $j = 0,1,2, \dots, g$ and $\alpha_i^0 = 0, i = 1,2, \dots, h$
- For $m = 0,1,2,$ do the following until $max(|\alpha_i^m - \alpha_i^{m+1}|) < \varepsilon$

$$\beta_j^{m+1} = u_{-j} + \frac{\sum_{i=1}^{\infty} \alpha_i^m \beta_{i+j}^m}{1 - \beta_0^m} \quad for\, j = 0, 1, \dots, g \tag{4.6}$$

$$\alpha_j^{m+1} = u_j + \frac{\sum_{i=1}^{\infty} \alpha_{i+j}^m \beta_i^m}{1 - \beta_0^m} \quad for\, j = 1, 2, \dots, h \tag{4.7}$$

Now $\vec{z}$ is computed as follows:

$$z_0' = 1 - \sum_{i=1}^{h} \frac{\alpha_i}{1 - \beta_0} \tag{4.8}$$

For the probability of $z' = 0$ and

$$z_i' = \sum_{j=1}^{h} \frac{z'_{i-j} \alpha_j}{1 - \beta_0} \tag{4.9}$$

In which z'corresponds to the distribution of the customer waiting time. In the inventory context, this corresponds to the inventory level before the satisfaction of demand. In order to obtain the inventory level after the satisfaction of demand, i.e. the total throughput time, we need to convolve the deficit before demand with the distribution of demand:

$$\vec{z} = \vec{z}' * \vec{d} \tag{4.10}$$

In the case of order lead times of n periods, we need to compute the n-fold convolution $\vec{d}^{n*}$of the deficit with demand

$$\vec{z} = \vec{z} \quad * \vec{d}^{n*} \tag{4.11}$$

The procedure is summarized in Figure 4.6.

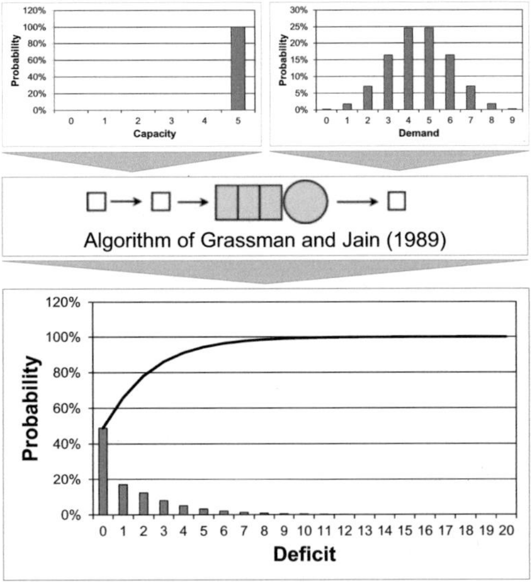

Figure 4.6: Calculating the probability distribution of the deficit by means of a queuing system

Given the probability distribution of the deficit, we can calculate the system's performance figures, such as the average physical inventory for a given number of kanbans N_K. Therefore, we use equation (4.3) to transform the probability distribution of the deficit into the probability distribution if the physical inventory. Given this distribution, we can simply calculate its expected value to determine the average physical inventory.

The probability vector of the inventory consists of the following two parts:

$$\overrightarrow{I_{phys}} = \begin{pmatrix} i_0 \\ i_n \end{pmatrix} \tag{4.12}$$

In equation 3.6, i_0 denotes the probability that inventory is empty, i_n denotes all other states of i except being empty. They can be computed from the probability distribution of the deficit.

$$i_o = P(Z \geq N_K) = \sum_{j=N_K}^{z_{max}} z_j \tag{4.13}$$

$$i_n = P(Z < N_K) = z_{N_K - n} \; \forall \, n = i_1, \dots, i_{max} \tag{4.14}$$

The equations can be explained as follows. Any time the deficit is greater than the number of kanbans N_K, inventory is zero and the systems is currently in backlog. Therefore, in order to calculate the probability of inventory being exactly zero, we need to add up the all the probabilities of the system states in which the deficit z is greater than the number of kanbans. In case the deficit is smaller than the number of kanbans, the probability i being exactly n $P(i = n)$ corresponds to the probability of z being,$N_K - n$ i.e. $P(z = N_K - n)$.

$$E(I_{phys}) = \sum_{I_{phys,min}}^{I_{phys,max}} P(I_{phys} = i) * i \tag{4.15}$$

As pointed out by Meyer (2015), in practice we usually have delivery patterns with varying interarrival times which are not synchronous with demand. This means, demand and replenishment occur at different times. To incorporate this observation in our model, we need to apply stochastic lead time models as given for example by Tempelmeier (2011)

The problem is illustrated in Figure 4.7. In the figure, the system state is evaluated ten times. In each of the ten periods, there is a demand for products. A capacity for replenishment is only available in four periods. In addition, there is the aggregate demand, which is the sum of demand between two replenishments. The pattern is assumed as repetitive, therefore after period 9 we sum up the period demands of period 10, 1, 2 and 3 to end up with an aggregate demand of 4 in period 3.

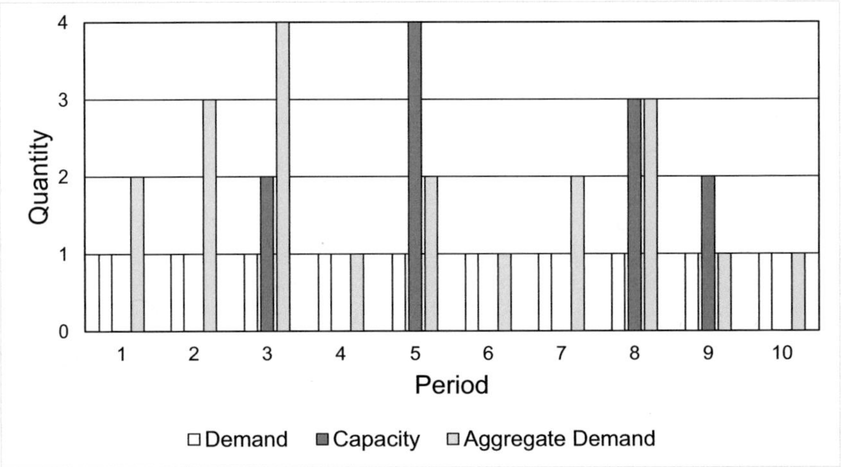

Figure 4.7: Leveling pattern with variable interarrival time

If we simply transform the pattern given in Figure 4.7 into a probability distribution, the probability that we have a capacity of 0 is 60%. This would imply that with a probability of 0.6^{10}, we would have no capacity during the whole leveling horizon. However, since we know that we definitely have capacity in periods 3, 5, 8 and 8, the model is invalid. Therefore we change our modeling approach by evaluating the system state only in periods in which there is capacity. In order to do that, we extract the capacity and aggregate demand bars from Figure 4.7 and only evaluate the system state in periods 3, 5, 8 and 9. This is depicted in Figure 4.8.

By evaluating the system state only in periods in which we have capacity, the probabilities of the capacity distribution changes. The probability of the capacity being zero is now 0%. With a probability of 50%, we have an available capacity of 2 units, with a probability of 25% we have 3 units and with a probability of 25% 4 units. Since we change the capacity distribution, we also need to adapt the demand distribution to reflect the instants of system state evaluation. This is why we need to calculate the aggregate demand.

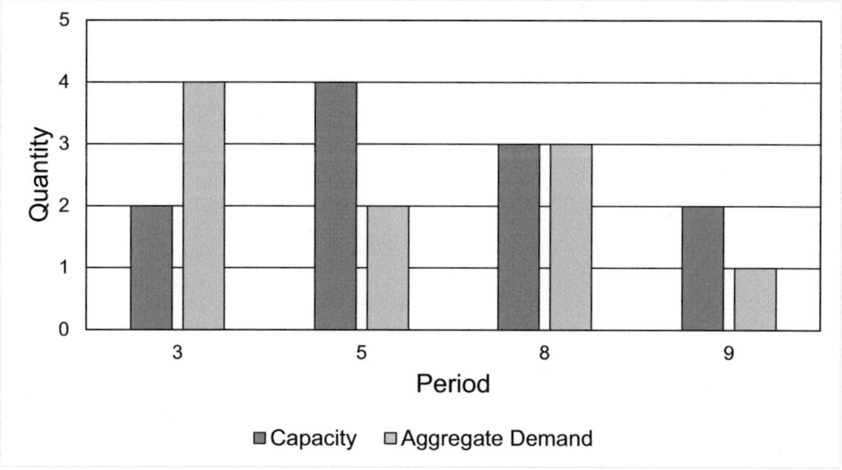

Figure 4.8: Leveling pattern with constant interarrival time and aggregate demand

To calculate the aggregate demand, we first have to calculate the interarrival time distribution given by the respective delivery pattern. This interarrival time distribution corresponds to the lead time distribution L in a discrete time stochastic lead time inventory model (see Figure 4.9). Now, we calculate the discrete demand distribution for each possible outcome l of the interarrival time by $(l-1)$-fold convolution. Afterwards, we weight the demand distribution of each interarrival time outcome with its respective probability and sum it up.

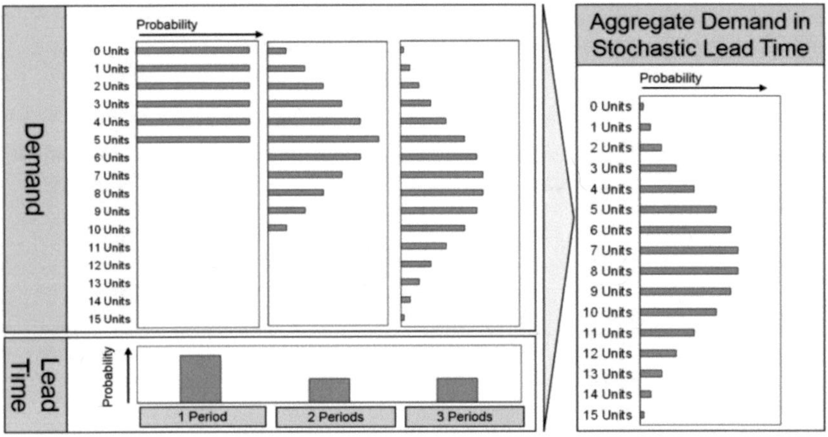

Figure 4.9: Application of stochastic lead time models to model delivery patterns with varying interarrival time

The calculations for each possible outcome d of the aggregate demand distribution which are outlined above can be computed by the following formula:

$$P(D \leq d) = \sum_{l_{min}}^{l_{max}} P(D \leq d \,|L = l) * P(L = l) \tag{4.16}$$

4.3.2 Capacity Reservation

The previous section presented a description of the inventory behavior in case of variable consumption but leveled replenishment. The object of investigation was a single part number. In this chapter we look at all parts number which share a common resource and analyze the decision regarding which part number can be ordered on which day and in which quantity. Therefore, we create a pattern which tries to order as level as possible in a vehicle which is as small as possible.

An example of this decision problem is illustrated in Table 4.3. In the example, the leveling period or EPEI encompasses four days and three different parts need to be transported. For each part i, a total number of units per EPEI b_i is specified. Moreover, each part has a certain unit capacity requirement w_i.

As stated above, the goal is to find a pattern which is as level as possible with respect to the ordered quantities, therefore creating a smooth flow and keeping inventory low. Our leveling pattern must further satisfy two types of constraints. First, the total quantity procured over the leveling period must be equal to the total required quantity. Second, on each day the total capacity requirement of our order must be smaller than the total capacity of the vehicle.

Table 4.3: Optimization problem of finding a leveled delivery pattern

Part No.	Capacity per Unit	t=1	t=2	t=3	t=4	Total
i=1	w_1	x_{11}	x_{12}	x_{13}	x_{14}	b_1
i=2	w_2	x_{21}	x_{22}	x_{23}	x_{24}	b_2
i=3	w_3	x_{31}	x_{32}	x_{33}	x_{34}	b_3
Total		C_v	C_v	C_v	C_v	

Table 4.4 summarizes the parameters and variables we use for our optimization model. In addition to the parameters that were mentioned above, we define the set of periods in the planning horizon $\mathcal{T}$. The total number of periods in the planning horizon is given by N_T. Since we perform a tour on each day in our planning horizon, N_T also corresponds to the number of tours in the planning horizon. Moreover, we consider the set $\mathcal{V}$ of vehicles. Each vehicle v has capacity of C_v. The binary decision variable θ_v indicates, which vehicle is used to perform the tours.

Table 4.4: Overview of parameters and variables of the optimization model

Parameters, Sets and Variables	
b_i	Total required quantity of part i
C_v	Capacity of vehicle v
N_T	Number of periods in planning horizon
p_i	Penalty cost of uneven orders of part i
$\mathcal{T}$	Set of number of periods in planning horizon $\{1, \dots, N_T\}$
θ_v	Binary decision variable: Use/do not use vehicle v
$\mathcal{V}$	Set of vehicles with different sizes/payloads
w_i	Capacity requirement of part i
x_{it}	Quantity of part i on day t, decision variable
y_{it}	Deviation from mean demand of part i on day t, decision variable

The problem can be solved by means of mathematical optimization. The formulation is stated below.

$$Minimize \sum_i \sum_t y_{it}^2 \cdot p_i + M \cdot \sum_{v \in \mathcal{V}} \theta_v \cdot C_v \tag{4.17}$$

subject to

$$\sum_t x_{it} = b_i \ \forall i \in \mathcal{P} \tag{4.18}$$

$$\sum_i w_i \cdot x_{it} \leq \sum_{v \in \mathcal{V}} \theta_v \cdot C_v \ \forall t \in \mathcal{T} \tag{4.19}$$

$$\sum_{v \in \mathcal{V}} \theta_v = 1 \tag{4.20}$$

$$x_{it} - y_{it} = \frac{b_i}{N_T} \forall i \in \mathcal{P}, t \in \mathcal{T} \tag{4.21}$$

$$x_{it} \in \mathbb{Z}^n \tag{4.22}$$

$$\theta_v \in \{0,1\} \tag{4.23}$$

Constraint (4.18) ensures that for each part i of the set of parts $\mathcal{P}$, the total quantity which is reserved is equal to the total demand of the whole leveling period. Constraint (4.19) ensures that the total capacity of the parts transported the day is smaller than the vehicle capacity. Constraint (4.20) limits the number of vehicles to 1. Constraint (4.22) assures that we only order whole shipment units.

Constraint (4.21), the leveling constraint, requires that each order quantity x_{it} is equal to the mean demand of the period plus the excess quantity y_{ij} which is penalized in the objective function (4.17). Put simply, the constraint tries to force us to order the same quantity of b_i/N_T, i.e. the mean demand, every day. Because of the vehicle capacity restriction this is not possible. Since y_{ij} is part of the objective function, we ensure that each time we deviate from the mean, this is punished with the factor p_i in the objective function. Moreover, the objective function chooses the minimal vehicle size from set of vehicles which is sufficient to perform the transport job.

The penalty costs enable us to priorize products regarding their need for leveled orders. As weighting factor we could for instance choose the part price or part storage requirements. Since ordering in an unleveled pattern leads to higher buffer inventory, we only want low-priced parts with low capacity requirements to be ordered uneven.

4.4 System Design

The preceding section presented mathematical models of two subsystems of the heijunka leveled transport logistics system. This section explains how the system can be designed. We first give an overview of the steps we need to follow. Afterwards, the subsections explain each step of the design process in detail.

Figure 4.10 summarizes the major steps in the design process. At first, each supplier needs to be assigned a transport concept. Afterwards, each supplier which was assigned the milk run concept needs to be assigned a tour. For each tour, a leveling pattern is calculated. Given the leveling pattern, the necessary buffer inventory can be calculated. The result of the planning steps is a Plan for Every Part (PFEP) for transport logistics.

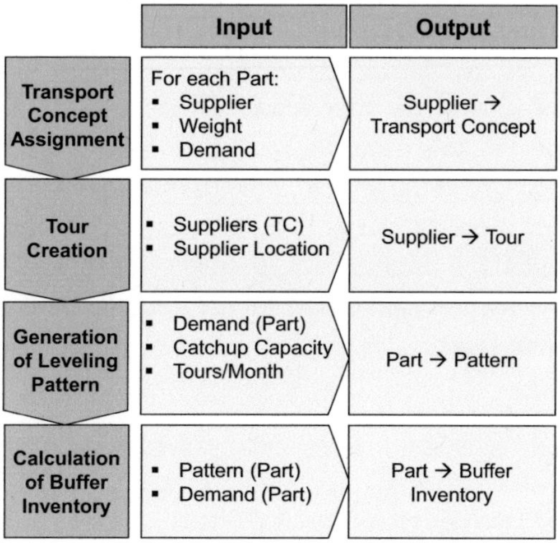

Figure 4.10: System design - creating a Plan for Every Part

The PFEP contains the following information for each part of our supply network:

- Part Number
- Part Supplier
- Part Weight/Capacity Requirements
- Average Part Consumption and Variability of Consumption
- Transport Concept
- Leveling Pattern
- Number of Kanbans

In the following sections, we describe each of the steps from above in detail. The first step is the assignment of a transport concept.

4.4.1 Transport Concept Assignment

Following Meyer (2014), the first decision to take is the assignment of the transport concept. For this decision, we follow VDA 5010 with some slight modifications. Because of the leveled replenishment policy, the stability of the required transport capacity can always be regarded as high. Moreover, the optimization model (cf. section 4.3.1) ensures that load consolidation is possible. This is why, as displayed in Figure 4.11, the milk run is applicable for a wider range of cases.

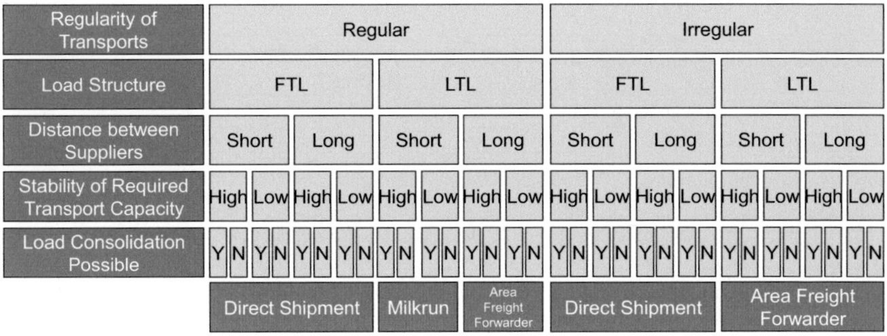

Figure 4.11: VDA 5010 - Adapted for the case of a leveled replenishment policy

First, all suppliers need to be classified regarding their delivery frequency and load structure. To quantitatively grasp the load structure, we calculate the expected capacity per day. That is, for each supplier we multiply the expected demand D_i with the required unit capacity w_i and take the sum over all parts i:

$$\sum_{i=1}^{n} E(D_i) \cdot w_i \tag{4.24}$$

According to the principles of lean manufacturing, it is most desirable to order in the highest possible frequency in order to keep the buffer inventory, needed to ensure the provision of parts during two deliveries, as low as possible. If there are suppliers from which we order as much (or as heavy/voluminous) parts, such that we could order at both an acceptable delivery frequency (e.g. 1/day or 1/week) and an acceptable utilization of transport capacity, we assign the supplier to the concept of direct shipment.

If the loads are too small and delivery frequencies are low, the transports are conducted by an area freight forwarder. If the loads are too small and the delivery frequencies are sufficiently high, the next step is to check whether there are further suppliers in the proximity, so that we can build tours. If there are no suppliers in the proximity, the supplier needs to be assigned to the AFF concept. If there are, the supplier is assigned to the milk run concept.

4.4.2 Creation of Tours

After the clustering of suppliers and the transport concept assignment, we have to create feasible tours for all milk run tours. In case of direct shipment, the transport simply consists of a source (the supplier) and a sink (the receiving plant). Therefore there is no need creating tours. If the supplier is assigned the AFF concept, the tours are planned ad-hoc by the freight forwarder.

Feasibility is restricted by the driving time regulations by law. Due to regulations of the European Union, the daily driving time must not exceed nine hours. It can be extended to 10 hours at most twice per week (see (European Parliament and Council 2006)). Therefore, the number of suppliers which can be included in one tour is limited.

The tours can be created by a variety of algorithms, e.g. different versions of the vehicle routing problem. In this case, our goal is only to create feasible tours. Therefore we use sweep algorithm because it is a simple algorithm and can be used to generate feasible tours.

Figure 4.12 illustrates how the sweep algorithm can be applied. Gillett and Miller (1974) describe the steps as follows:

- Initialize by setting the "sweep line", e.g. at 6 o clock
- Move the sweep line clock wise
- Add first supplier to the tour
- Move the sweep line and
    - If the tour stays feasible, add the next supplier to the tour.
    - Else, close tour and start a new tour at supplier
- Stop, if all suppliers are assigned to a tour

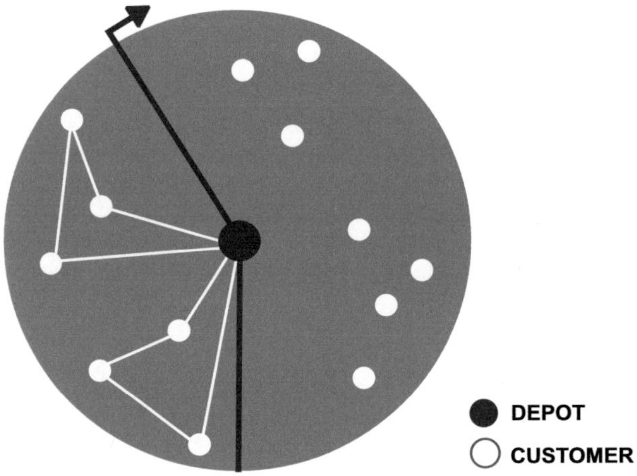

Figure 4.12: The sweep algorithm

For checking the feasibility, it is important to take into account that the tour starts and ends at the receiving plant and that every supplier has a service time. That is we calculate the total duration of the tour which includes the 'next' supplier by adding travel times (incl. the way back to the depot) and service times. If the time is smaller than 9h, we check if the tour is still feasible when we add the next supplier.

4.4.3 Calculation of a Leveling Pattern

Given the tours that were created according to the preceding section, we can create a leveled pattern for each tour. This is why in this chapter we explore how the optimization model presented in section 4.3.1 can be used to calculate the heijunka pattern of each part. We first investigate the simple case of one staged direct milk runs, i.e. there is no consolidation point between the suppliers and the receiving plant. In the second part we investigate how we can model transport systems which consist of a pre- and a main leg, i.e. there is a consolidation point between the supplier and the receiving plant.

4.4.3.1 Direct Tours

For the calculation of the leveled pattern, we employ the optimization model from section 4.3.1 in two different ways. These ways depend on the Strategies that are followed. The first step is to calculate the total quantity per part b_i which needs to be procured over the leveling horizon.

To calculate b_i, we need to define how much catch-up capacity CUC_i for each part i we want ensure. As already noted in section 2.4, variability needs to be buffered by some combination of capacity, inventory, or time to buffer this variability. Choosing a high catchup capacity needs less buffer inventory. However, if we want a lower catchup capacity, we will need more buffer inventory to achieve a certain service level.

If $E(D_i)$ denotes the expected demand per day in the leveling period and N_D stands for the number of days in the leveling period, then $E(D_i) \cdot N_D$ denotes the expected demand in the leveling period. Dividing the expected demand in the leveling period by the number of units we reserve b_i yields the planned utilization of the reserved capacity. The catch-up capacity is defined as difference between 1 and the planned utilization. This can be expressed as follows:

$$CUC_i = 1 - \frac{E(D_i) \cdot N_D}{b_i} \tag{4.25}$$

Note that b_i can only take integer values, since it is only permitted to order whole shipment units. Furthermore, the expected demand also must be calculated in shipment units.

In the leveling period, we want to reserve the minimum capacity for each part. However, in order to avoid too high buffer inventories caused by a lack of catchup capacity, we calculate the number of capacity slots, which we reserve on a truck for part i as follows:

The number of units that need to be reserved for each part b_i can be calculated by multiplying the expected demand $E(D_i)$ with the number of days in the leveling period N_D and rounding it up to next integer value. If this value does not violate our minimum catchup capacity restriction, this is the number of capacity slots we reserve. However, if it violates the restriction, we reserve the minimum number of capacity slots, which is in accordance with our minimum catchup capacity restriction. That is, we divide the demand during the leveling horizon $E(D_i) \cdot N_T$ by $(1 - CUC_i)$ and round up to the next integer.

$$b_i = \begin{cases} \left\lceil \dfrac{E(D_i) \cdot N_D}{1 - CUC_i} \right\rceil \forall i \in \mathcal{P} & if\ CUC_i > 1 - \dfrac{E(D_i) \cdot N_D}{\lceil E(D_i) \cdot N_D \rceil} \\ \lceil E(D_i) \cdot N_D \rceil \forall i \in \mathcal{P} & else \end{cases} \tag{4.26}$$

When we calculated the total quantities per part for the whole leveling horizon, we need to take a decision regarding the vehicle capacity. We can essentially follow two different Strategies:

- First set a desired vehicle capacity and then calculate the required frequency
- First set a desired frequency and then calculate the required vehicle size.

Both of these Strategies can be mapped to our optimization model by adapting and providing the input data. The first step is to calculate the total capacity TC which is required over the whole leveling horizon. It can be calculated as by summing up the capacity requirements of all the parts i which are picked up in one tour:

$$TC = \sum_{i \in \mathcal{P}} b_i \cdot w_i \tag{4.27}$$

If we know the total capacity required in the leveling period, we can now calculate the number of tours N_T or the vehicle capacity C_V, depending on the Strategy we follow:

$$N_T = \left\lceil \frac{TC}{C_V} \right\rceil \tag{4.28}$$

$$C_V = \left\lceil \frac{TC}{N_T} \right\rceil \tag{4.29}$$

With these as input, we can use the optimization model described in section 4.3.1 to determine a leveled delivery pattern.

4.4.3.2 Pre- and Main Leg Tours

In practice, it is not ensured that all suppliers are in direct proximity to the receiving plant and can be served by a milkrun. Usually, this condition only holds for a small part of suppliers. In these cases, it makes sense to procure materials from these suppliers by a two stage transport concept, i.e. to split the transport chain into a pre- and a main leg by introducing a consolidation point.

In a two stage transport concept, the pre- leg can be organized like a usual milkrun. The only difference is that the goods are not transported directly to the receiving plant but to a consolidation point, e.g. a cross-docking center. At this consolidation point, parcels from different tours are sorted regarding their destination plant. The subsequent main leg transport brings them to its destination.

The integration of a consolidation center creates further potential. In a production network, different receiving plants sometimes need the same part from the same supplier. They can now be picked up by the same pre-leg milkrun as one parcel, which is then split up at the consolidation center. This offers the potential of increasing the transport frequency and thus lowering the inbound inventory at the receiving plants. This is illustrated in Figure 4.13 for the case of six suppliers served by two tours. The vehicles pick up goods at these suppliers and bring them to the cross-docking center. There, they are sorted and brought to their destination plant by three main-legs.

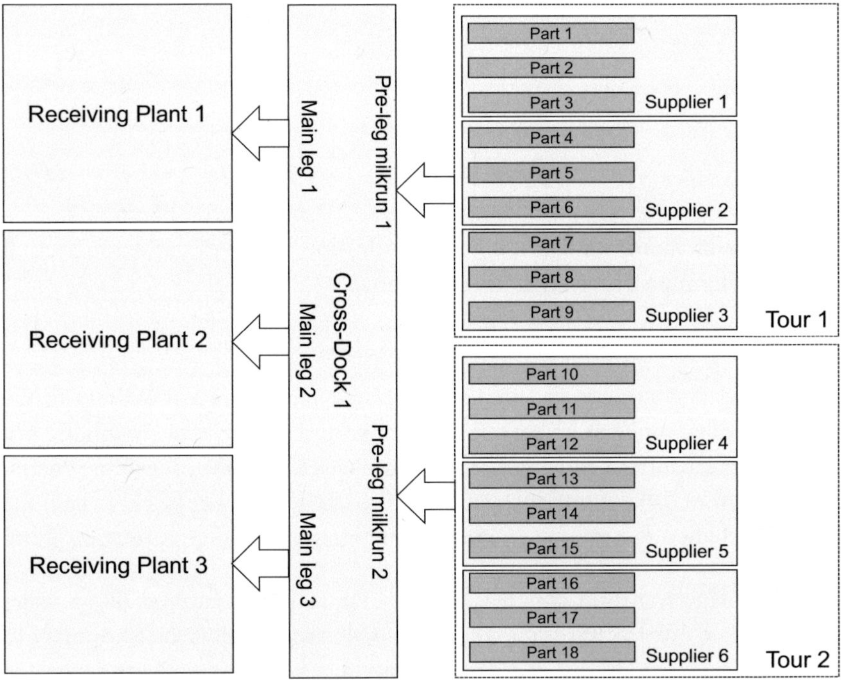

Figure 4.13: Elements of a two stage transport system

In order to calculate delivery schedule for such a system, we need to take into account further requirements

Generally, excessive buffering in the cross-docking centers is not possible. Therefore the frequencies of the pre-leg and main-leg must be aligned

Moreover, if multiple receiving plants need the same part from the same supplier, this can be modeled as n different parts for n receiving plants or one part for n receiving plants. In the first case, the system design is trivial and no different to the methods presented in this work. In the second case, the design gets more complicated. If multiple receiving plants share a common capacity slot and on one day their demand exceeds the reserved capacity, a priority rule is required. This priority rule yields a decision regarding which supplier gets how many parts and thus has an effect on the capacity distribution of the inbound supermarket of each supplier. Up to now, there are no inventory models for heijunka-leveling with these priority rules.[1]

After having decided on the frequency of the main-leg, we can calculate the delivery patterns for one tour. Therefore, we first calculate three sub-delivery patterns for the parts of all three receiving plants. The overall delivery schedule is calculated by the superposition of the three sub-delivery patterns.

4.4.4 Calculation of Buffer Inventory

Having calculated the delivery pattern, we are now able to calculate the buffer inventory needed to achieve a desired service level. We assume that the distribution of demand for parts is known. In practice, it could for example be created from the orders of the past three months or export the delivery schedule for the upcoming three months from the ERP system.

Figure 4.14 summarizes the steps that need to be performed to determine the buffer inventory. First, the probability distribution of the capacity needs to be derived from the delivery pattern which was created with our optimization

[1] In a production environment, this problems corresponds to heijunka leveling on a family level. That is the part family would be "the part" and all "members" of the family would be the different receiving plants.

model in section 4.3.1. Since demand and capacity are not synchronous, they need to be transformed to two processes on the same discrete time scale.

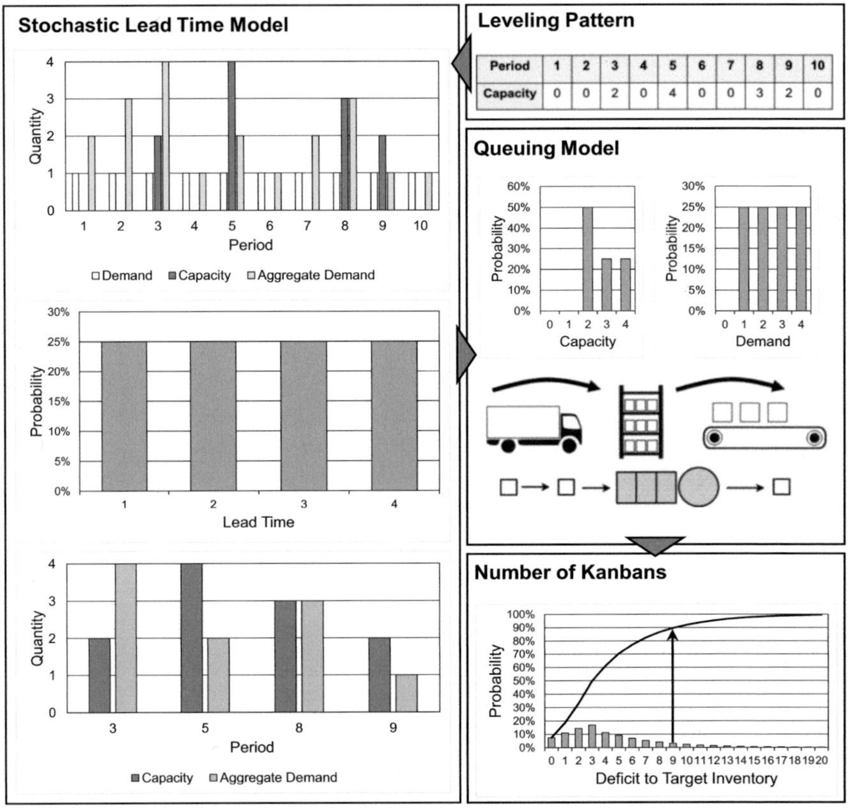

Figure 4.14: Interfaces between optimization model and inventory model

For the transformation, we only evaluate the system state in periods with a capacity greater than zero. To transform the demand to the same time scale as the capacity, the demand between two replenishment periods needs to be summed up. This can be accomplished by calculating the inter-arrival time distribution of the capacity and using the stochastic lead time model (cf. section

4.3.1) to calculate the aggregate demand in stochastic lead time. In each period on the new time scale, the system state changes by the superposition of the capacity process and the aggregate demand process. The capacity and aggregate demand are the input factors of our GG1 queuing model.

The queuing model yields the probability distribution of the deficit, as displayed in the last step of Figure 4.14. Given the probability distribution of the deficit, we can also compute the cumulative probability distribution of the deficit. If we would like our inventory to be sufficient in 99% of all periods we choose the number of kanbans at which the cumulative distribution function of the deficit reaches the 99% line. This means, in only one percent of all periods the deficit is larger than our number of kanbans, i.e. the inventory is zero.

5 Evaluating the Effectiveness of Leveling

We must have had 99 percent of the game. It was the other three percent that cost us the match.
(Ruud Gullit, Dutch football player)

The preceding chapter described how the concept of heijunka leveling can be transferred to transport logistics as a measure to increase the stability of the required transport capacity. We built mathematical models to understand the system behavior and showed how these can be employed to design the system. On this basis, we want to evaluate the effectiveness of heijunka leveling as a measure to stabilize transport logistics systems by means of a simulation study in this section

In the first part of this chapter we describe the structure and the functionality of the simulation model we employ for the effectiveness evaluation. Subsequently we present the design of experiments, i.e. according to which logic we create the sample data and how we derive our scenarios. Afterwards, we present the results regarding the effectiveness in stabilizing the required transport capacity. Moreover, we present the results regarding the effectiveness on stabilizing the replenishment orders on the individual part level.

5.1 Description of Simulation Model

We suggest a hierarchical agent-based modeling approach to evaluate the effectiveness of heijunka leveled material supply as described in section 4. Its structure is depicted in Figure 5.1

The top level agents are the suppliers, the logistics service provider (i.e. the truck) and the receiving plant. Each top level agent contains several sub agents. The receiving plant consists of production processes, which consume raw material and inventory models of certain parts. The parts and the production processes are linked by a bill of material. This enables us to model amplification effects (e.g. one produced unit needs two parts) or correlation effects (one product needs two different parts).

Each inventory agent is modeled by an inventory policy (cf. section 3.2.1). For our experiments, the inventory agents can either behave like a kanban system, which is modeled as an (r, S) policy with a maximum backlog of 0, or like a heijunka system. The inventory is reviewed daily, i.e. $r = 1\ day$. The heijunka system is also modeled as an inventory policy consisting of the parameters $(r, S, \tilde{q})$. In this policy, $\tilde{q}$ denotes the maximum order quantity of a certain period. This corresponds to the number of capacity slots that is reserved in the heijunka board (cf. section 4). Put simply, the policy corresponds to an (r, S) policy with the following additions:

- if $S - I_{pos} <= q_{max}$, order $\tilde{q} = S - I_{pos}$ units
- if $S - I_{pos} > q_{max}$, order $\tilde{q} = q_{max}$ units

Again, we allow no backlog, i.e. we encounter lost sales.

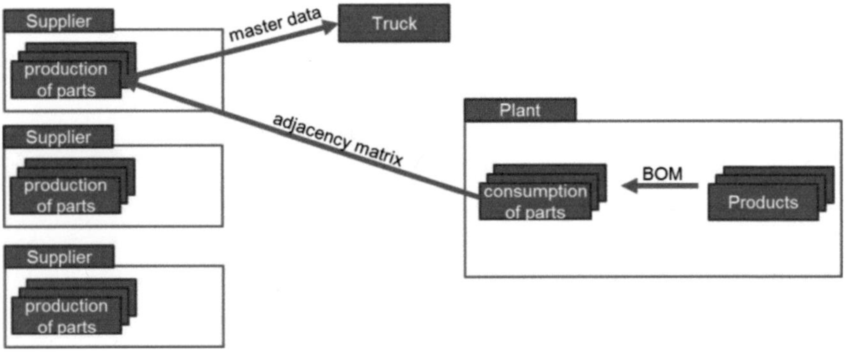

Figure 5.1: Structure of the agent-based simulation model to investigate leveling effectiveness.

In case of leveled replenishments following the heijunka principle, we can also employ a delivery pattern which is aligned with the vehicle capacity (see section 4.4.3). That is, the maximum size of the orders that can be placed is not the same for each day but varies in a 20-day recurring pattern.

In addition to the receiving plant agent, the model contains multiple supplier agents. Each supplier agent contains a population of part production agents which provide the material that is requested according to the replenishment policy described above. Each part consumption agent is uniquely connected to one part production agent.

The third type of agents in the model is the truck agent. Its task is to pick up materials at the suppliers and deliver it to the receiving plant. The tours are triggered according to a fixed schedule, i.e. it behaves like a milk run (see section 3.1.3). Moreover, the model contains master data for each part to calculate the capacity requirements, e.g. product weight or product volume.

The performance figures we measure in order to evaluate the effectiveness of the heijunka-leveled replenishment policy are displayed in Table 5.1. We collect both time series data regarding demand and replenishment orders. On the part level we measure the coefficient of variation of the quantities that were ordered and replenished. The reason is that on the part level, we want to measure how our policy influences the bullwhip effect. This is usually done by relating the cv^2 of orders to the cv^2 of demand. On the transport level, we want the total sum of our orders to fit in a certain truck size with a specified statistical safety. Therefore we measure the 99%-quantile of the required capacity.

Table 5.1: Performance figures, points of measurement and dimensions measured in the simulation model

	Part Level	Transport Level
Demand (before leveling)	cv_D^2 $(quantity)$	$q_{0.99}$ $(capacity)$
Orders (after leveling)	cv_O^2 $(quantity)$	$q_{0.99}$ $(capacity)$

As shown in Table 5.2, we collect data at the level of individual parts and at the transport level, i.e. the sum of the capacity requirements of all individual products. This enables us to distinguish if the measure is effective on only one of these two levels or on both levels. We distinguish between the needed capacity, i.e. the weight or volume, and the quantity. In order to increase the utilization of our truck, it is necessary that the total required capacity is stabilized. Moreover, if the quantities of the replenishment orders of our different parts are stabilized, this is beneficial for the suppliers (see also in section 4.1).

The logic of our measurements is further explained in Figure 5.2. The production consumes raw material of each part i. This is denoted by d_i. This consumption lowers the inventory level of each part i. Depending on the selected replenishment policy, replenishment orders are placed at the supplier. The truck picks up the material and delivers it to the receiving plant. This results in an increase of the inventory. In order to evaluate the relative performance of the leveled and the unleveled case, we collect data before leveling, i.e. the consumption of parts by the production, as well as after leveling, i.e. the replenishment orders placed at the suppliers.

To measure the effectiveness on the **transport level**, we calculate the total weight of all parts that were consumed. Further, we calculate the total weight of the replenishment orders that were placed at the suppliers. This is the weight that is transported by the truck.

On **part level** we compare the demand d_i and and the orders o_i of each part i individually.

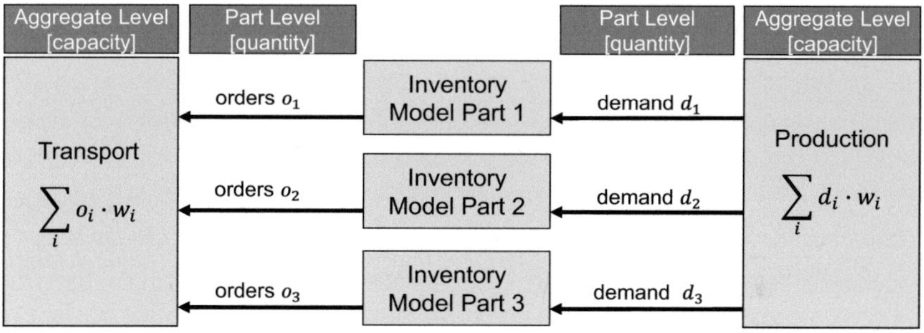

Figure 5.2: Measuring points in the simulation model

The two performance measures we collect are the squared coefficient of variation and the 99% quantile $q_{0.99}$. The cv² indicates whether the replenishment policy amplifies or dampens the demand fluctuations, i.e. whether the bullwhip effect is present or not. The 99%-quantile q_{99} is our measure of the required capacity. In practice we would want our truck to have a capacity sufficient to transport all the goods with a certain statistical safety, e.g. 99%. If we are able to decrease this 99% quantile, the result is that we can use a smaller truck and achieve the same average throughput and service level (see section 3.3) because the available capacity is better utilized.

If the leveling is effective, both the cv² and the 99%-quantile q_{99} will be decreased by the leveled replenishment policy.

5.2 Design of Experiments

On the basis of the simulation model described in the preceding section, we will now describe the experiments we conduct with the model. In the first part of the section we describe the sample data and the logic we followed to create it. In the second part of the section. We describe the different parameter combinations that can be created from the sample data.

5.2.1 Generation of Sample Data

The factors determining the required transport capacity each day are the quantity per part and the required unit capacity per part in weight or volume units. The quantity is a Poisson-distributed random variable with a certain mean, whereas the required unit capacity is a constant. Therefore, the required capacity is a linear transformation of the random demand. Each part is characterized by a combination of these two parameters.

In practice, both quantity and weight distributions are usually following a so-called Pareto distribution. That is, the quantities or weights are unevenly distributed (e.g. 20% of parts correspond to 80% of total quantity). This means, there are high runners and low runners as well as heavy parts and light-weight parts. (Alicke 2005).

A measure to quantify inequality of samples is the Gini coefficient (Gini 1921). It can be calculated from the Lorenz curve, as depicted in Figure 5.3, by dividing the area A between the line of equality and the Lorenz curve by the total area under the line of equality, A+B. Originally, the figure comes from economics to measure the distribution of wealth among inhabitants of a country. A Gini coefficient of G=0 corresponds to perfect equality, i.e. everybody is equally wealthy. G=1 stands for perfect inequality: One person owns everything, the rest of the population owns nothing.

The principle can be also transferred to our simulation experiment. In case of the mean demand, G=0 means that all parts have the same mean demand, G=1 that all the demand is for one parts and the remaining parts have a demand of 0. The same applies for the unit weight: in case of G=0, all parts have the same unit weight, in case of G=1, all parts but one part have a unit weight of 0.

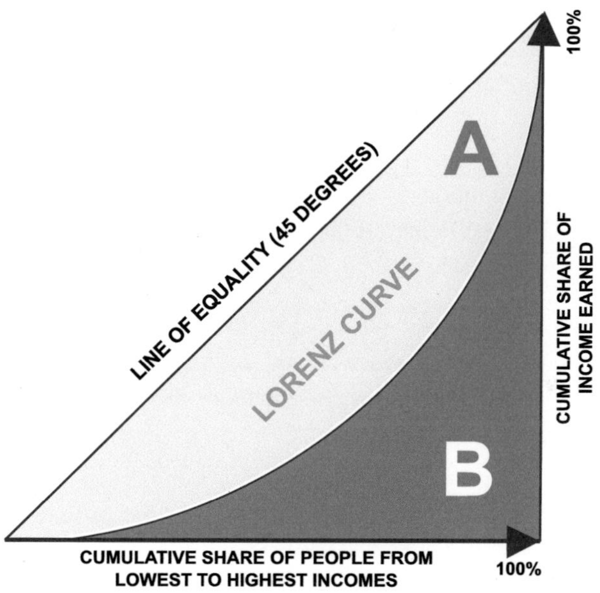

Figure 5.3: The Lorenz curve as basis for the calculation of the Gini coefficient (cf. Mankiw 2015)

The distributions for the G=0.25, G=0.50 and G=0.75 were created by using the so-called zeta distribution (cf. Bronstein 2016). It is the discrete equivalent of the Pareto distribution, which yields a discrete distribution for a certain exponent s and i elements. The index i corresponds to a part i. Its values can be computed as follows.

$$f\ (i) = \frac{i^{-s}}{\zeta(s)} \tag{5.1}$$

In (5.1), $\zeta(s)$ denotes the Riemann zeta function. It is given by

$$\zeta(s) = \sum_{n=1}^{\infty} \frac{1}{n^s} \tag{5.2}$$

To obtain the distribution of mean demands, we iteratively varied s and calculated the Gini coefficient of the distribution. We then picked probability distributions with G=0.25, G=0.50, G=0.75.

In our simulation experiment, we consider 150 parts. Each of these parts has a Poisson-distributed demand with a certain mean λ_i and a capacity requirement, e.g. weight w_i. Each $\lambda_i$is given by equation (5.1). All part demands are independent from one another.

$$P_\lambda(k) = \frac{\lambda_i^k}{k!} \cdot e^{-\lambda_i} \tag{5.3}$$

The total sum of the mean demands is the same for each parameter combination. We only vary the Gini coefficient that is how the total demand is distributed among parts. The same applies for the unit weight: The total sum of unit weights is the same for all parameter combinations, only the distribution is varied.

We investigate two weight-mean demand scenarios: The heavy high runners (i.e. light low runners) and the heavy low runners (i.e. light high runners). These combinations are depicted in Figure 5.4.

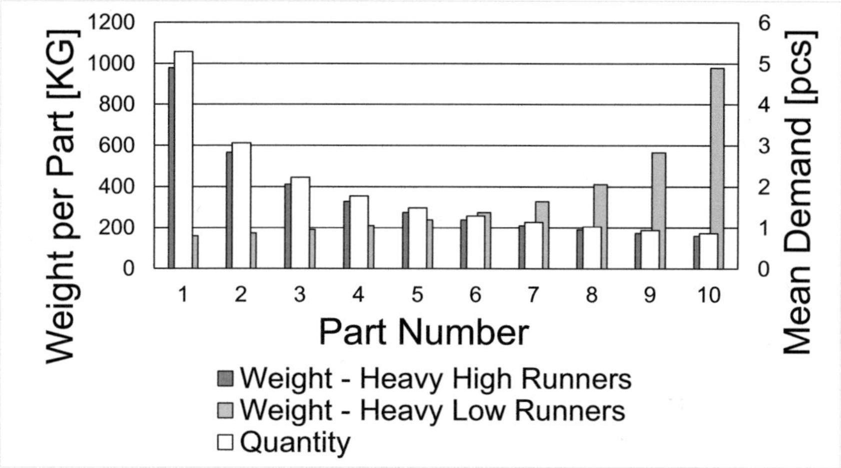

Figure 5.4: Design of experiments - heavy high runner and heavy low runner scenarios

In practice, of course, many different (quantity, weight) tupels are possible. In our simulation study, we cannot investigate all possible combinations. However, all of these combinations lie in between the two extreme cases of our simulation scenarios.

5.2.2 Simulations Scenarios

As stated in section 5.2.1, we differentiate between different levels of inequality regarding demand and weight. For both dimensions, Gini coefficients of 0, 0.25, 0.5, 0.75 and 1 are investigated. We further consider two different scenarios of $(G_{Weight}, G_{Quantity})$ tupels: heavy high runners and heavy low runners. Putting it all together, this yields the 50 parameter combinations depicted in Table 5.2.

Table 5.2: Parameter combinations and simulation scenarios

		$G_{Quantity}$				
		0.00	0.25	0.50	0.75	1.00
G_{Weight}	0.00	Scenario 1: Heavy High Runners				
	0.25					
	0.50					
	0.75					
	1.00					
	0.00	Scenario 2: Heavy Low Runners				
	0.25					
	0.50					
	0.75					
	1.00					

These parameter combinations of inequality include some special cases. In case of $G_{Quantity} = 1$, the demand for all parts but one is zero. Hence, this is the one product case. The corresponding column shows the behavior of increasing weight. The one product case gives us an indication regarding how the system behaves in case of correlated demand. If all products were perfectly correlated, the system behaves just like the one product case with the weight of the single product being equal to the sum of all part weights. Without a leveled replenishment, this is a kind of worst case scenario. In practice, we usually operate somewhere in between perfect correlation and perfect independence. That is, some parts are correlated and some parts are independent.

Another special case is the one product case of $G_{Weight} = 1$. In this case, the weight of all parts except for one is zero. That is, they do not have an effect on the total weight. If we vary $G_{Quantity}$ in case of $G_{Weight} = 1$, the mean of our Poisson demand increases and the coefficient of variation decreases. All mean demands and unit weights that were created can be reviewed in the appendix.

These scenarios are investigated with two different order policies, i.e. the kanban policy and the heijunka policy. In respect to the heijunka policy, we further distinguish between a leveling period of a 1-day recurring pattern and a 20-day recurring pattern.

5.3 Effectiveness on Transport Level

The first kind of analysis is our investigation on the transport level (cf. Figure 5.2). In this investigation, we want to determine how leveled replenishment of parts effect the total transport capacity required to perform the transportation task.

The first subsection gives an overview by analyzing the simulation results on the scenario level as depicted in Table 5.2. Afterwards we go into detail by presenting the results on the parameter combination level.

5.3.1 Overview on Scenario Level

The goal of this section is to present an overview of the simulation results of the different parameter combinations on an aggregate level. Therefore, we calculate the mean of our performance measures for each scenario.

As pointed out in chapter 1, stability is the probability of a process outcome being within a desired target range. In our case, the process outcome is the total weight on truck which is observed each day and the target state is defined by the total truck capacity. We keep the total throughput constant. Therefore, if we need a smaller truck, the replenishment policy has increased the stability of the weight on truck.

The stability figures we analyze are the 99% quantile and the cv^2 of the total capacity required to fulfill the replenishment orders each day. The replenishment policies of our investigation are heijunka leveling with a 1-day (HE1d) and a 20-day (HE20d) recurring pattern. To measure the effectiveness, we put $q_{0.99,d}$ in proportion to $q_{0.99,o}$ for both policies. In order for our leveling policy to be considered as effective in increasing the stability of the total weight on truck, this ratio must be smaller than one. A ratio equal to one means no effect, a ratio greater than one means a detrimental effect, i.e. the stability has been lowered.

On the transport level, we calculate the total capacity needed in each period of our simulation run. It can be expressed as follows:

$$TC_{o,t} = \sum_{i \in \mathcal{P}} o_{i,t} \cdot w_i \tag{5.4}$$

$$TC_{d,t} = \sum_{i \in \mathcal{P}} d_{i,t} \cdot w_i \tag{5.5}$$

For each parameter combination, we calculate the ratio of the 99% quantiles QR by dividing the 99% quantile of the total required capacity of the orders $q_{0.99,o}(TC)$ by the 99% quantile of the total required capacity of the demand $q_{0.99,d}(TC)$.

$$QR_r = \frac{q_{0.99,o}(TC)}{q_{0.99,d}(TC)} \; \forall \, r \in \{HE1d, HE20d\} \tag{5.6}$$

For the aggregation on basis of the scenarios, we calculate the mean value of all $N_{PC} = 25$ parameter combinations belonging to a certain scenario. The set of the heavy high runner parameter combinations is called $\mathcal{S}_{HHR}$, the set of heavy low runners is called $\mathcal{S}_{HLR}$. We refer to the elements of both sets by the index s. Therefore, we can calculate their mean values as follows:

$$\overline{QR_{r,s}} = \frac{1}{N_{PC}} \sum_{s=1}^{N_{PC}} QR_{r,s} \; \forall \, r \in \{HE1d, HE20d\}, \forall \, s \in \{\mathcal{S}_{HHR}, \mathcal{S}_{HLR}\} \tag{5.7}$$

Figure 5.5 shows that for all cases that were investigated, the quantile ratio QR is smaller than one. That is, the stability was increased by heijunka leveling. The same average transport throughput could have been achieved with less capacity. Since in a kanban policy, the replenishment orders are equal to the demand, $q_{0.99,d}$ is equal to $q_{0.99,o}$ in the heijunka case. Therefore, operating the kanban system with a heijunka replenishment policy would have resulted in reductions of the required transport capacity between 17% and 28%.

We further note that for both scenarios, the longer delivery pattern was more effective than the shorter pattern. This is plausible. We need to reserve enough capacity for each part and can only place orders of discrete sizes. Therefore we need to round up the mean demand to the next integer to reserve a sufficient amount of capacity. The longer the pattern gets, the more accurate our rounded reserved capacity corresponds to the actual mean demand and the less excess capacity we reserve. With a two day recurring pattern, we can round up to a precision of 0.5 units. With a five day recurring pattern, we can round up to a precision of 0.2 units. With a twenty day recurring pattern, we can round up to a precision of 0.05 units.

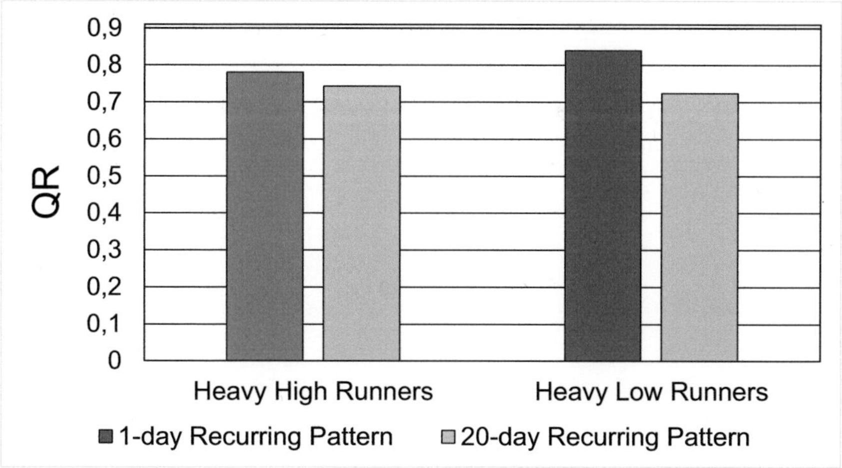

Figure 5.5: QR in case of heavy high runners and heavy low runners for different leveling horizons

Moreover, we observe that the difference between the 1-day recurring pattern and the 20-day recurring pattern is higher in the heavy low runner cases. The reason is again related to the fact that we reserve excess capacity by rounding up the mean demand to the next integer. The low runners have a low demand which is often smaller than one unit per day. If we reserve capacity of one unit per day, there is excess capacity reserved which diminishes the effect of leveling. The relative amount of excess capacity created due to the integer rounding is higher in case of a low demand. Since the low runners are the heavy parts, their proportion of the total weight on truck is relatively high despite the low quantity in which they are needed.

Figure 5.6 shows a similar analysis, the difference being that we investigate the cv^2 of the weight on truck instead of the 99% quantile. On the y-axis, the squared coefficient of variation of the orders $cv^2\ (O)$ is set in proportion to the squared coefficient of variation of demand $cv^2\ (D)$.

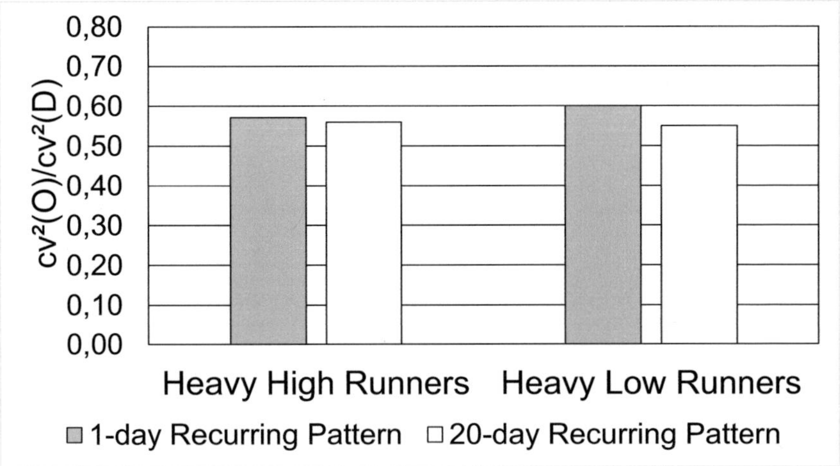

Figure 5.6: Relation of order-cv^2 to demand cv^2 for different scenarios

Again, we note that all the ratios are smaller than 1. This means that the orders were more stable than the demand, i.e. the leveled replenishment policy was effective. For both scenarios, the effectiveness of leveling is higher in case of a twenty day recurring pattern. The results are qualitatively in accordance with the results reported above for the $q_{0.99}$.

5.3.2 Analysis of Parameter Combinations

After presenting the overview of the results on the scenario level in the preceding section, we will now present the results for each parameter combination. We first investigate the case of heavy high runners. Afterwards the case of heavy low runners is investigated.

5.3.2.1 Heavy High Runners

The first class of scenarios we analyze are the group of heavy high runners. That is, the high runners are also the heavy parts and low runners are the lightweight parts

Table 5.3 shows the ratio of the 99%-quantiles of the weight of demand to the weight on truck QR for both a leveling pattern of one day and a leveling pattern of 20 days for all of the scenario's 25 parameter configurations (cf. equation 5.6).

In all observed parameter combinations, QR is smaller than 1. That is the $q_{0.99,o}$ is smaller than $q_{0.99,d}$. For increasing inequality of weight, the ratio QR decreases, which means that the effectiveness of the leveling policy increases. This is true for all columns but the one product case of $G_{Quantity} = 1$, in which, of course, the leveling effect is the same for all cases. We cannot observe any linear relationships between the Gini coefficient and the QR ratio.

Table 5.3: Ratio of 99%-quantiles in case of heavy high runners

		$G_{Quantity}$									
		1-day Recurring Pattern					20-day Recurring Pattern				
		0.00	0.25	0.50	0.75	1.00	0.00	0.25	0.50	0.75	1.00
G_{Weight}	0.00	0.94	0.94	0.94	0.91	0.84	0.93	0.93	0.91	0.90	0.84
	0.25	0.94	0.86	0.83	0.83	0.84	0.91	0.84	0.83	0.82	0.84
	0.50	0.90	0.74	0.76	0.76	0.84	0.60	0.61	0.70	0.74	0.84
	0.75	0.73	0.61	0.69	0.69	0.84	0.50	0.56	0.64	0.72	0.84
	1.00	0.50	0.52	0.55	0. 67	0.84	0.50	0.52	0.55	0.67	0.84

In the one-product case for $G_{Quantity} = 1$, QR is constant for all weight scenarios. This is trivial since the maximum order quantity is the same for all weight scenarios. We also observe that for $G_{Quantity} = 0$ and $G_{Weight} = 0$, QR is closest to one. That is, the relative benefit of a heijunka controlled kanban system compared to a regular kanban system decreases. The reason is the pooling effect, which is most pronounced in case of many different parts with independent demand and equal weight. As we stated in section 5.2.2, it is less pronounced if we have only one part or a correlation between part demands.

Figure 5.7 displays the frequency distribution of the weight on truck for different replenishment policies in case of the parameter combination $G_{Quantity} = G_{Weight} = 0.5$. We investigate plain kanban replenishment (in red), heijunka replenishment with a delivery pattern with a length of one day (in yellow) and heijunka replenishment with a delivery pattern of a length of 20 days (in green).

All frequency distributions have the same mean as depicted by the black dashed line. The vertical lines represent the 99% quantiles of the different replenishment policies.

Our base case scenario is the regular kanban replenishment. If we look at the red bars, i.e. the weight on truck in case of a kanban controlled replenishment, we notice that the distribution roughly follows the bell-shape of a Gaussian normal distribution.

This is plausible. In a kanban system, we order exactly the quantity that has been consumed before, i.e. our orders correspond to the demand in the preceding period. That is, our order quantities also follow a Poisson distribution. The weight on the truck is the sum of the weights of all the parts which have been ordered in a certain period. The quantities are independent random variables and the total weights are linear dependent on the quantities. According to the central limit theorem[1], they must be normally distributed.

We notice that the green and yellow vertical lines are left to the red vertical line. This means that for a statistical safety of 99% we would have needed less transport capacity to transport goods from the supplier to the receiving plant. Since in case of kanban replenishment, the distribution of replenishment orders is equal to the demand distribution[2], the red distribution also shows the demand distribution. Since $q_{99,Kanban} > q_{99,Heijunka}$ for the 1-day and the 20-day pattern, we conclude that heijunka leveling was effective in stabilizing the weight on truck in the case of heavy high runners.

In Figure 5.7 the green vertical line is left to the yellow vertical line. The reason for this are rounding effects that occur due to the fact that we can only order discrete quantities, i.e. no partly filled shipment units. Therefore the quantities we reserve on the truck for the part numbers must also be discrete. By rounding up to the next integer, we reserve excess capacity. The effect is more pronounced for low runners than for high runners. This is quite straightforward, as this simple example shows: Imagine a part with a mean demand of 0.1 units/day. In case of a 1-day recurring pattern, we would need to reserve one quantity slot for this part every day, yielding a catchup capacity

[1] According to the central limit theorem, a sum of independent random variables approaches a normal distribution, if the length of the sum approaches infinity.

[2] This is not entirely true. If there is no physical inventory left, lost sales occur in kanban system. For a sufficiently high number of kanbans, this effect is negligible.

of 90%. In case of a part with a mean demand of 10.1 units/day, we would reserve 11 units. This results in a catchup capacity of $1 - 10.1/11 = 8.2$ %. In case of a 20-day pattern, we could order 2 units per 20 days in the first case and 202 units per 20 days in the second case, leaving no catchup capacity and thus maximizing the effect of leveling.

We further notice that $q_{99,Kanban} - q_{99,HE1d} > q_{99,HE1d} - q_{99,HE20d}$. This can be explained by the observation that we are looking at a heavy high runner scenario. That is, the high-weight parts are also the high runners. Thus, their impact on the total weight is higher than the impact of the low runners. Since the rounding errors that lead to catchup capacity are smaller for these high runners, heijunka leveling with a pattern length of 1 day is almost as effective as heijunka leveling with a pattern length of 20 days.

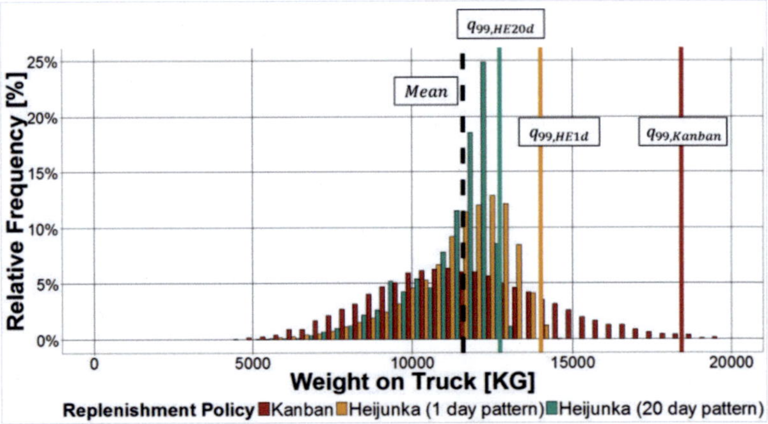

Figure 5.7: Frequency distribution of weight on truck for different replenishment policies in case of heavy high runners

In this section we investigated the effectiveness of a heijunka-leveled replenishment policy for the case of heavy high runners, i.e. the parts with a throughput in terms of quantity units per unit also have high unit weight. In the next section, we investigate a different scenario: now the parts with a low throughput are those with a high unit weight.

5.3.2.2 Heavy Low Runners

The second class of scenarios we analyze is the group of heavy low runners. That is, the low runners are also the heavy parts and high runners are the lightweight parts.

Table 5.4 summarizes the simulation results of this scenario. Again, we calculated the QR according to equation (5.6) for both a leveling pattern of one day and a leveling pattern of 20 days.

Again, all cases QR is smaller than 1, which means that $q_{0.99,o}$ in case of heijunka-leveled replenishment is smaller than $q_{0.99,o}$ in case of the unleveled kanban replenishment. We cannot observe any linear relationships between the Gini coefficient and QR. For increasing inequality of weight, QR decreases, which means that the effectiveness of the leveling policy increases. This observation holds for all columns but the one product case, in which the leveling effect is the same for all parameter combinations.

Table 5.4: Ratio of 99%-quantiles in case of heavy low runners

		$G_{Quantity}$									
		1-day Recurring Pattern					20-day Recurring Pattern				
		0.00	0.25	0.50	0.75	1.00	0.00	0.25	0.50	0.75	1.00
G_{Weight}	0.00	0.96	0.95	0.93	0.91	0.84	0.88	0.87	0.87	0.86	0.84
	0.25	0.94	0.94	0.93	0.92	0.84	0.72	0.75	0.79	0.83	0.84
	0.50	0.86	0.91	0.93	0.92	0.84	0.60	0.64	0.71	0.75	0.84
	0.75	0.75	0.82	0.93	0.91	0.84	0.55	0.59	0.61	0.71	0.84
	1.00	0.50	0.55	0.60	0.64	-	0.50	0.55	0.60	0.64	-

Figure 5.8 again displays the frequency distribution of the weight on truck for the kanban policy, heijunka with a 1-day recurring leveling pattern and heijunka with a 20-day recurring leveling pattern for the parameter combination $G_{Quantity} = G_{Weight} = 0.5$.

We observe that $q_{99,Kanban} > q_{99,HE1d} > q_{99,HE20d}$. Therefore we conclude that heijunka leveling is also effective in the case of heavy low runners.

Again, the frequency distribution of the weight on truck roughly follows a Gaussian bell curve. The skewness is a little higher than in the case of heavy low runners making it appear more like a Poisson distribution. The reason is that in the present case, the low runners are the heavy parts. The quantities ordered in each period are Poisson distributed and the skewness of the Poisson distribution increases with decreasing mean. Therefore, the total weight on truck is more skew in case of heavy low runners than in the case of heavy high runners.

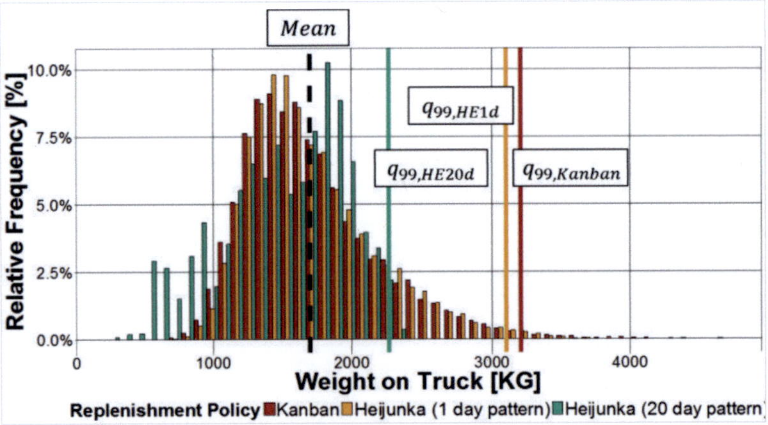

Figure 5.8: Frequency distribution of weight on truck for different replenishment policies in case of heavy low runners

In contrast to the case of heavy high runners, this time $q_{99,Kanban} - q_{99,HE1d} < q_{99,HE1d} - q_{99,HE20d}$. As we explored in the preceding section, the effect of rounding up to the next integer when reserving truck capacities leads to more catchup capacity for low runners than for high runners. This catchup capacity diminishes the effectiveness of leveling. That is, the heavy parts, which have a higher contribution to the total weight than the light parts, are only leveled to a small extent which is why the heijunka system with a 1-day pattern almost behaves like a kanban system.

By introducing a 20-day pattern, the effectiveness of heijunka leveling is increased. Whereas $q_{99,\mathrm{HE1d}}$ is 96.6% of $q_{99,\mathrm{KA}}$, $q_{99,HE20d}$ is only 70.5% of $q_{99,\mathrm{KA}}$. The reason is that the longer the pattern, the smaller the rounding-induced catchup-capacity.

5.3.3 Data Collection

As stated in the preceding section, we investigate 50 different parameter combinations that can be classified according to the two scenarios of heavy high runners and heavy low runners.

For each parameter combination, we conduct 10 simulation runs (replications) with a random seed. The random number generator is only used to generate the fluctuating Poisson demand. The rest is entirely deterministic. One simulation run consists of 5,000 periods, i.e. 5,000 transports. Therefore, on transport level, we collect 50,000 data points per variable. On part level, we collect 50,000 data points per part. With 150 parts this yields 7,500,000 data points per variable.

We initialize the system with the analytically calculated average inventory for each part. That is, the number of kanbans that was calculated by the inventory model. Since we start the simulation in the steady state, there is no need for a warm-up phase and all the data points that are collected can be used for the evaluation of the experiments.

5.4 Effectiveness on Part Level

In the preceding section we analyzed the effectiveness of heijunka leveling on the transport level. In this section, we analyze its effectiveness on the individual part level. That is, for each part in our simulation study, we compare the demand and the respective replenishment orders. In contrast to the preceding section, the figures of our focus are the quantities, not the capacities (cf. Table 5.1).

In the first part of the section, we evaluate the patterns resulting from our optimization model regarding its fit with the minimum possible cv^2. In the second part, we evaluate the effectiveness based on the results of our simulation study.

5.4.1 Evaluation of Optimization Results

The first step of our evaluation on the part level is the evaluation of the solution quality of the results of our optimization model which is used to generate a leveling pattern in the case of a pattern length of 20 days. We first perform an analytical calculation to generate an inventory-minimal pattern which is possible under the assumption of unlimited capacity and compare to the solution yielded by the optimization model which is subject to capacity constraints.

In the theoretical optimum, we order parts in the smallest possible order lot sizes in a rhythm of constant interarrival times. Our optimization model, in contrast, tries to order in the smallest possible order sizes subject to a capacity constraint. Hence, if capacity is low, sometimes we are forced to order in lot sizes which deviate from the optimum minimal lot size.

Since we can only generate orders of discrete sizes and our leveling period consists of twenty days, we can calculate the minimal possible cv^2 as a function of the number of units that need to be ordered in total over the whole leveling period. The relationship is displayed in Figure 5.9.

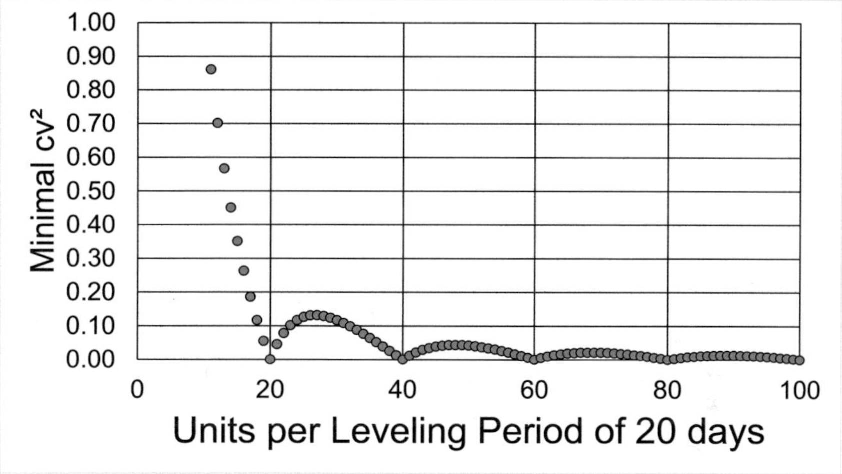

Figure 5.9: Calculation of minimal possible cv² as a function of the number of units per leveling period

We note that $cv^2 = 0$ is only possible if we order exactly 20 units per month or an integer factor of it. The maximum cv^2 depends on the length of the leveling period. In case of a 20 day delivery pattern, the maximum possible cv^2 is 20 which occurs in case of one order per month.

For each parameter combination, b_i is given and we can calculate the minimal possible cv². Since it only depends on $G_{Quantity}$ and is independent of G_{Weight}, we calculate the minimal possible cv^2 for 5 parameter combinations × 150 parts per parameter combination × 1 leveling pattern per part = 750 leveling patterns and compare it to the patterns generated by our optimization model. The results are displayed in Figure 5.10

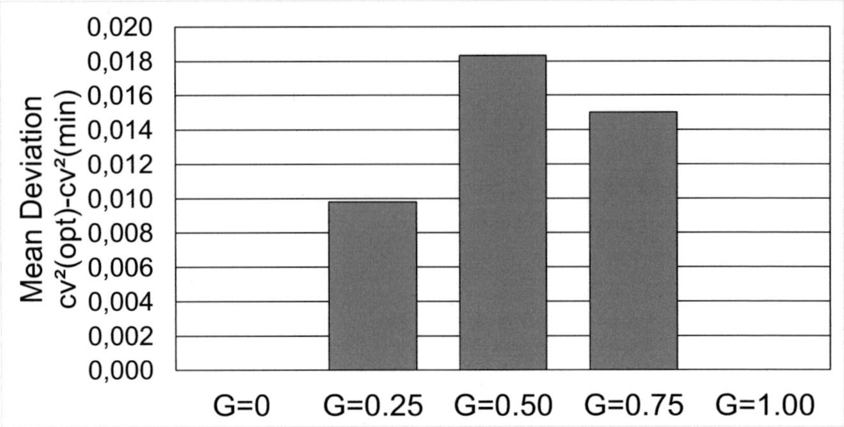

Figure 5.10: Mean deviation of cv^2 calculated by the optimization model from the minimal possible cv^2

The horizontal axis displays the different $G_{Quantity}$ of our simulation study. The vertical axis displays the mean deviation of the cv^2 of the pattern generated by our optimization model from the minimal possible cv^2. We note that in case of $G_{Quantity} = 0$ and $G_{Quantity} = 1$, the solution of the optimization model is equal to the optimal solution. The worst solution quality was achieved in case of $G_{Quantity} = 0.5$, where the deviation was 0.018.

In Figure 5.11 we calculated the percentage of parts in which the solution generated by our optimization model is equal to the optimal solution. In case of $G_{Quantity} = 0.5$, the case with the lowest solution quality, 5.3% of parts were assigned a pattern which was not equal to the optimal pattern.

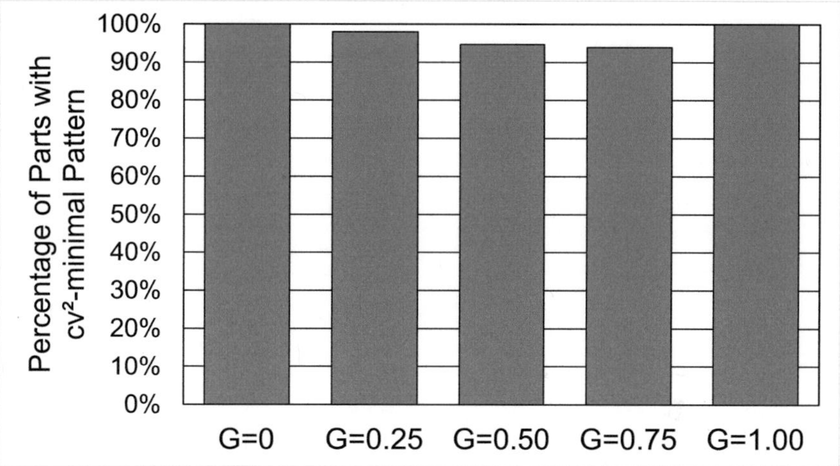

Figure 5.11: Percentage of parts procured with the lowest possible cv^2

The reason for this amplification of variability is an adverse superposition of parts in the leveling pattern that leads to capacity peaks in case of a less variable heijunka pattern. Since the capacity each day is restricted by a constraint in our simulation model and the evenness of the pattern is a variable in the objective function, the optimization leads to an uneven pattern on the part level for the sake of creating a pattern which is even regarding the daily total weight. That is, the solution is detrimental regarding the variability of orders on the individual part level but beneficial regarding the total weight on the transport level.

5.4.2 Evaluation of Simulation Results

In the preceding section we investigated the effect of leveled replenishment by comparing the replenishment pattern generated by the optimization model to a theoretical minimum. In this section, we explore, to what extent the variation of the replenishment orders has been decreased in comparison to the given demand.

For the evaluation, we analyze the time series of the requested quantity of each part D_i and the replenishment orders O_i. We compare orders and demand for each part in each of the 50 parameter combinations. As a quantitative figure to evaluate the effectiveness, we calculate the squared coefficients of variation of both demand and orders. In case of the demand, is defined as the quotient of the sample standard deviation of demand $s(D_i)$ and the mean demand $\overline{d_i}$. For the orders, it can be calculated likewise.

$$cv^2(D_i) = \left(\frac{s(D_i)}{\overline{d_i}}\right)^2 \quad \forall\, i \in \mathcal{P} \tag{5.8}$$

$$cv^2(O_i) = \left(\frac{s(O_i)}{\overline{o_i}}\right)^2 \quad \forall\, i \in \mathcal{P} \tag{5.9}$$

To measure the effect of leveling, we calculate the difference between $cv^2(D_i)$ and $cv^2(O_i)$ for each part and calculate the mean of all parts for each parameter combination j of the two sets of scenarios $\mathcal{S}_{HHR}$ and $\mathcal{S}_{HLR}$.

$$\Delta_{cv^2,i} = cv^2(D_i) - cv^2(O_i) \;\forall\, i \in \mathcal{P} \tag{5.10}$$

$$\overline{\Delta_{cv^2,j}} = \frac{1}{N_P}\sum_{i=1}^{N_P} \Delta_{cv^2,i} \quad \forall\, j \in \{\mathcal{S}_{HHR}, \mathcal{S}_{HLR}\} \tag{5.11}$$

In addition to the difference Δ_{cv^2}, we calculate the ratio r of the coefficients of variation of the orders and demand. It is a figure frequently used as a measure regarding the demand amplification of the bullwhip effect (Veit 2010). If this ratio is greater than one, the variability of demand has been amplified and our system is subject to the bullwhip effect. If the ratio is smaller than one, we reduced the bullwhip effect.

We calculate r_{cv^2} for each part i in each parameter combination j as follows:

$$r_{cv^2,i} = \frac{cv^2(O_i)}{cv^2(D_i)} \; \forall\, i \in \mathcal{P} \tag{5.12}$$

$$\overline{r_{cv^2,j}} = \frac{1}{N_P}\sum_{i=1}^{N_P} r_{cv^2,i} \; \forall\, j \in \{\mathcal{S}_{HHR}, \mathcal{S}_{HLR}\} \tag{5.13}$$

In Figure 5.12 we calculated the mean difference between the order cv² and demand cv² and aggregated the results by calculating the mean per scenario. In the figure, we distinguish between a leveling pattern with EPEI=1d and EPEI=20d. We note that for all two groups, the difference is positive, i.e. the leveled replenishment policy lowered the coefficient of variation.

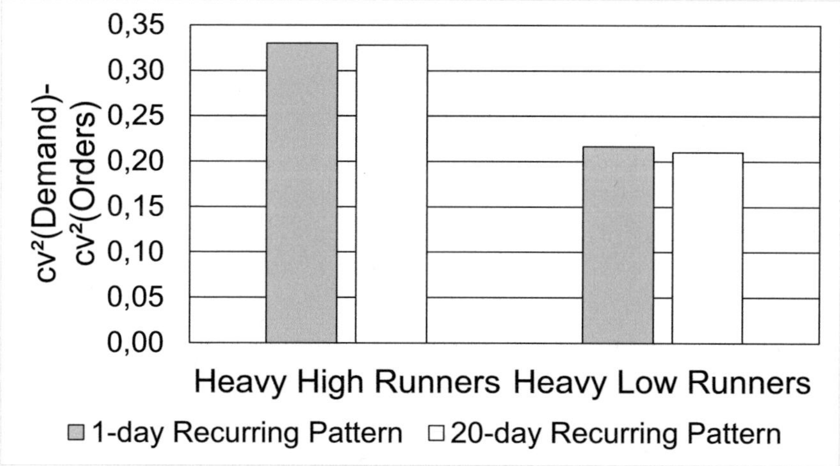

Figure 5.12: Difference between order cv² and demand cv² - mean of all parts per scenario

The figure also shows that on the individual level, both the 1-day pattern and the 20-day pattern are about equally effective, with a slight advantage for the 1-day pattern. The reason is the optimization procedure employed in case of the 20-day pattern. If it is required due to the capacity restrictions, variable interarrival times and varying order lot sizes are allowed. This reduces the effectiveness of leveling on the part level. In case of the 1-day pattern, each product can be ordered in the same fixed quantity every day. Therefore, there are no variable interarrival times and no varying order lot sizes.

Another observation from Figure 5.12 is that again, the effectiveness of leveling is lower in case of heavy low runners. The reason is that for the low runners, rounding up to the next integer leads to high amount of excess capacity diminishing the effectiveness of the leveled replenishment policy.

In Figure 5.13, we calculated the cv^2 ratio (cf. equation (5.13) for the two different scenarios. As expected, the ratio is equal to one in the kanban case, since variations in demand are simply passed on to the replenishment orders. In case of heijunka replenishment, the ratio is just below sixty percent. That is, on average, leveled replenishment reduced the variability by an amount of about 40%.

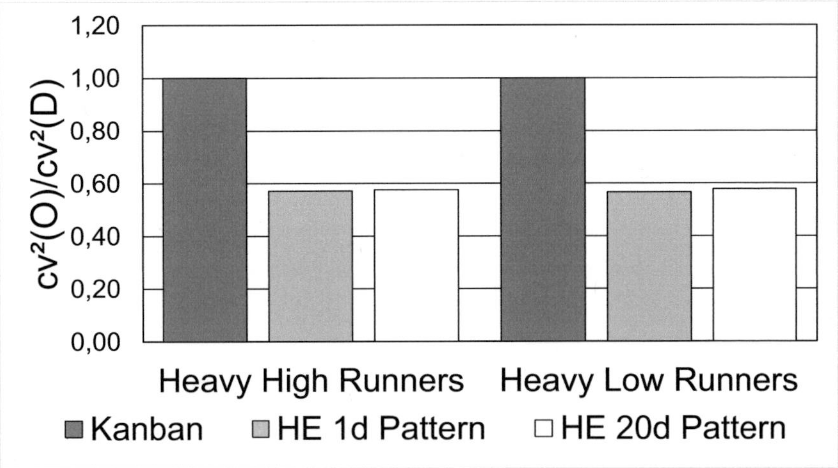

Figure 5.13: CV^2 ratio demand over orders – mean of all parts per scenario class

Figure 5.14 shows the frequency distribution of Δ_{cv^2} for all parts. We note that for heijunka leveling with a 1-day recurring pattern, the coefficient of variation for all parts in all parameter combinations could be reduced. This also was the case for most parts and parameter combinations with a 20-day recurring pattern. In about 96.7% of the observations, the variability of orders was lower than the variability of demand.

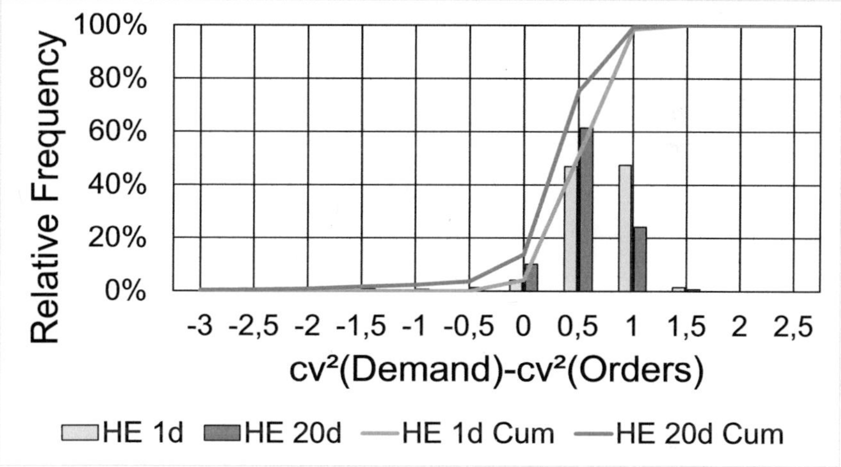

Figure 5.14: Frequency distribution: difference between order cv^2 and demand cv^2

In the remaining 4.3% of Figure 5.14, the coefficient of variation of the replenishment orders is higher than the coefficient of variation of demand. Figure 5.15 shows the scenario classes in which these amplifications of variability occurred. We note that the variability increases occurred almost evenly distributed in our two scenarios. This indicates, that the inequality of the weight per part is not a determining factor regarding the effectiveness on the part level.

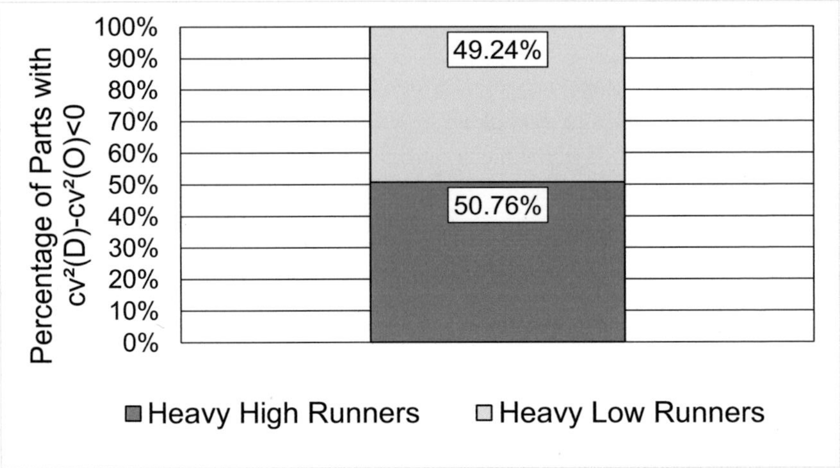

Figure 5.15: Percentage of parts where the delivery pattern led to an increase in variability

Table 5.5 displays the absolute frequency distribution of the observations of $\Delta_{cv^2} > 0$ in each of the 50 parameter combinations of our simulation study. We note that 97.3% of the observations occurred in the column $G_{Quantity} = 0.5$. The observation that the number of parts which were subject to an increase in variability are clustered in one column further supports our thesis that the inequality of weight is not an determining factor regarding the effectiveness of leveling on the part level.

The reason for this amplification of variability is, as pointed out above in section 5.4.1, adverse superposition of parts in the leveling pattern which leads to capacity peaks in case of a less variable heijunka pattern. Since the leveling pattern generated by the optimization model suffers from adverse superposition, this observation is also true for the simulation results.

Table 5.5: Absolute frequency distribution of parts subject to an increase in variability

			$G_{Quantity}$				
			0.00	0.25	0.50	0.75	1.00
G_{Weight}	Heavy High Runners	0.00	0	0	22	0	0
		0.25	0	1	23	0	0
		0.50	0	0	31	0	0
		0.75	0	1	26	0	0
		1.00	0	1	28	0	0
	Heavy Low Runners	0.00	0	0	27	0	0
		0.25	0	1	26	0	0
		0.50	0	1	25	0	0
		0.75	0	1	21	0	0
		1.00	0	1	26	0	0

The adverse superposition can be illustrated by the example of part 72 in the heavy high runner parameter combination $G_{Weight} = G_{Quantity} = 0.5$. In this parameter combination, the absolute deviation of the cv^2 from the minimum possible cv^2 amounts up 0.47. Both the solution of the optimization model and the optimal pattern are displayed in Table 5.6.

We note that the inventory minimal pattern is not perfectly smooth. In most periods, we can order one unit. In some periods, we order 2 units. These periods are split with an equal interarrival time of 4 periods. The optimization model, in contrast, created a solution which is even less smooth as can be seen by the higher cv^2. In many periods, 2 capacity slots are reserved while in other periods, the reserved capacity is 0.

Table 5.6: Leveling pattern of part 72 - optimization output vs inventory minimal pattern

	Period																				Sum	cv^2
	1	2	3	4	5	6	7	8	9	10	11	12	13	14	15	16	17	18	19	20		
Optimi-zation Output	2	2	0	0	0	0	2	2	2	0	0	2	1	2	2	2	2	2	2	0	25	0.6
Inventory Minimal	2	1	1	1	2	1	1	1	2	1	1	1	2	1	1	1	2	1	1	1	25	0.13

In order to create a more even pattern, we would need more transport capacity each day in order to buffer for the variability of the total weight. That is, a second kind of capacity buffer distinct from the catch-up capacity described in section 4.4.3. Again, the catch-up capacity on the transport level increases the transport costs but decreases the inventory costs. The reason is that the patterns generated by the optimization model in general order in smaller lot sizes and a less variable interarrival time.

5.5 Conclusion

In this chapter we investigated the effectiveness of the concept of heijunka leveling with respect to the stability of the required transport capacity and the stability of replenishment orders sent to the supplier. The results indicate that heijunka leveling is effective in stabilizing both the transport capacity and the replenishment orders.

On transport level, we found that in case of heavy high runners, heijunka leveling with a 1-day recurring pattern and a 20-day recurring pattern are about equally effective in stabilizing the required transport capacity with a slight advantage for the 20-day recurring pattern. In case of heavy low runners, the restriction of only ordering full shipment units leads to excess capacity, especially for products with a demand less than one shipment unit per day. This diminishes the effectiveness of leveling with respect to the total weight of the transport in the case of the 1-day recurring pattern.

For the case of heavy low runners we recommend to employ a 20-day (or longer) recurring pattern. It is more effective because it enables the planner to adjust the reserved capacity more exact to the actual demand. This leads to less excess capacity. The pattern increases the effectiveness of leveling and hence decreases the required transport capacity by a significant amount.

One more finding on the transport level was that the effectiveness of leveling increases with a higher inequality of the unit part weights. This observation can be explained by the pooling effect. The more evenly distributed and diversified the weight is among parts, the stronger the pooling effect. The more concentrated the weight is on one part, the closer we move to the one product case. In light of equation (3.1), the one-product case can be modeled as a perfect correlation, i.e. $\varrho = 1$.

We also observed that with increasing Gini coefficient of the mean demands, the effectiveness of leveling decreased in the one product case. The observation can be explained by the design of our experiments. Since the Gini coefficient of weight is one in the one product case, all weight is concentrated in part 1, the rest of parts have a weight of 0. For G=0, mean demand is evenly distributed among parts. With increasing Gini coefficient of mean demand, more mean demand is concentrated in part 1. In case of the Poisson distribution, the variability decreases with increasing mean. The lower the variability, the lower the potential effect of leveling. Therefore, in our experiment the increasing Gini coefficient of mean demands lead to a decrease in leveling effectiveness.

On the individual product level, both heijunka leveling with 1-day recurring pattern and a 20-day recurring pattern are effective in stabilizing the replenishment order quantities. The effectiveness is about equal with a slight advantage on 1-day recurring pattern. The reason is the optimization model which calculates the pattern in case of the 20-day recurring pattern. The objective function tries to order the smallest possible amount of each part number each day with respect to the capacity constraint of the truck. However, in case of adverse superposition this constraint leads to orders bigger than desired and hence a high coefficient of variation. In contrast, in the 1-day recurring, pattern, each day the same quantity of each part can be ordered. This leads to a decreased stability of replenishment order lot sizes in case of the 1-day recurring pattern.

We discovered that adverse superposition of parts can be an inhibitor to leveling. In these cases, a pattern with even order lot sizes still leads to an excessive amount of variation of the total capacity. Therefore, the optimization model creates a pattern with less variation regarding the total capacity at the expense of a higher variation on the part level. If the pattern needs to be smooth on the part level, for example due strategic restrictions or hard constraints regarding the maximum inventory, we can provide more transport capacity to buffer this variation. Please note that due to the law of variability buffering, the provision of an extra transport capacity buffer leads to smoother delivery patterns at the expense of vehicle utilization.

Based on these results, we conclude that heijunka leveling is an effective measure for stabilizing both the required transport capacity and the replenishment order quantities. The variability is, however, not eliminated by the Design for Stability. It is merely shifted from the transport capacity (and replenishment order quantity) dimension to the inventory dimension (cf. section 2.4). Therefore, the transport capacity buffer can be reduced while the inventory buffer needs to be increased. This leads to a trade-off, i.e. the costs for transport capacity decrease whereas the inventory costs increase.

In order to determine whether it makes sense for practitioners to employ heijunka leveling for their materials supply, it must be more efficient than their status quo materials supply. That is, the sum of the inventory and transport costs with heijunka leveling must be smaller than the status quo inventory and transport costs. Therefore, in the next section, we will create a mathematical model to evaluate the trade-off between inventory and capacity and show how to determine an efficient point of operation.

6 Evaluating the Efficiency of Leveling

Aw, people can come up with statistics to prove anything, Kent. 14 percent of all people know that.
(Homer Simpson, TV character)

In the preceding chapter we concluded that heijunka is effective in stabilizing the required transport capacity by performing an extensive simulation study on a versatile set of sample data. We found that heijunka leveling is an effective measure to reduce the variability of the required transport capacity. That enables us to use a truck with a smaller payload to achieve a given throughput and a required service level. Following the law of variability buffering, we now need a higher inventory buffer because variability was not eliminated but only shifted from one dimension to another. We saved buffer capacity at the expense of a higher buffer inventory. To determine an optimal buffer allocation as an efficient point of operation, we need to find a trade-off between these two conflicting objectives

In this chapter, we investigate the conditions that need to hold in order to make stabilizing the required transport capacity by employing heijunka leveling also an efficient measure. We first give an introduction about the basics of efficiency. Afterwards we investigate how the kanban and the heijunka system are linked by the catch-up capacity. Subsequently, we build an analytical model which enables us to calculate the system's total cost as a function of the selected buffer allocation. Based on that, we show how we can calculate an optimum buffer allocation, both in a numerical and an analytical way.

6.1 Basics of Efficiency

According to Pareto, efficiency is a state of allocation of resources from which it is impossible to reallocate so as to make any performance criterion better off without making one performance criterion worse off. That is, a so-called Pareto improvement, i.e. a change in allocation, which improves one criterion without making another criterion worse off, is not possible. If the current allocation allows for a Pareto improvement, it is called Pareto inefficient. In this case, a so-called deadweight loss or allocative inefficiency occurs (Varian 2014).

In heijunka leveling, these performance criterions are the cost of transportation and the cost of inventory. Both figures depend on the catch-up capacity which is reserved for the parts in excess of their required capacity. If there is a higher catch-up capacity, the cost of transportation rise since the transport cost are a linear function of transport capacity. The higher the capacity buffer, the lower the required inventory buffer. That is, the costs of inventory decrease. The minimum is at the point where the increase in cost of transportation is equal to the increase in cost of inventory. That is, the marginal cost of both components are equal. This minimum of the total cost function is also called the Pareto efficient buffer allocation.

The purpose of the catch-up capacity is to ensure that not all of the variability is buffered by inventory, but there remains a capacity buffer. Since there is a trade-off between buffering the variability with capacity and buffering it with inventory, the optimum is between the two extremes of an all-capacity or all-inventory buffer. The position of the optimum depends on the factor costs of capacity and inventory.

Before we determine the optimum catch-up capacity in the following sections of this chapter, we want to explore the relation between a kanban system and the heijunka leveled kanban system. The regular kanban system is in fact a special case of the heijunka leveled kanban system and both are linked by the catch-up capacity. This can be illustrated by a simple example, as depicted in Figure 6.1.

The figure displays the reorder quantity as a function of the current deficit to the target inventory. The curve is plotted for different amounts of reserved capacity. In a kanban system, the reorder quantity corresponds to the current deficit to the target inventory. It is represented by the linear slope of *Reorder Quantity = Deficit to Target Inventory*. In a heijunka system, the order size in each period is limited to a certain number of units. If the current deficit is smaller than the reserved capacity, we order exactly the deficit, i.e. there is no difference to the kanban system. If the deficit is greater than the reserved capacity, we only order the reserved capacity.

The more capacity we reserve, the more the heijunka system behaves like a kanban system. The reason is that heijunka is essentially a limitation of the reorder capacity. If the reserved capacity is equal to the number of kanbans in the loop, both systems behave the same because there is no limitation any more. With the increase in reserved capacity, the effectiveness of leveling increases but we need less inventory to buffer against the variations. The lower the reserved capacity, the stronger is the effectiveness of leveling but the more buffer inventory we need.

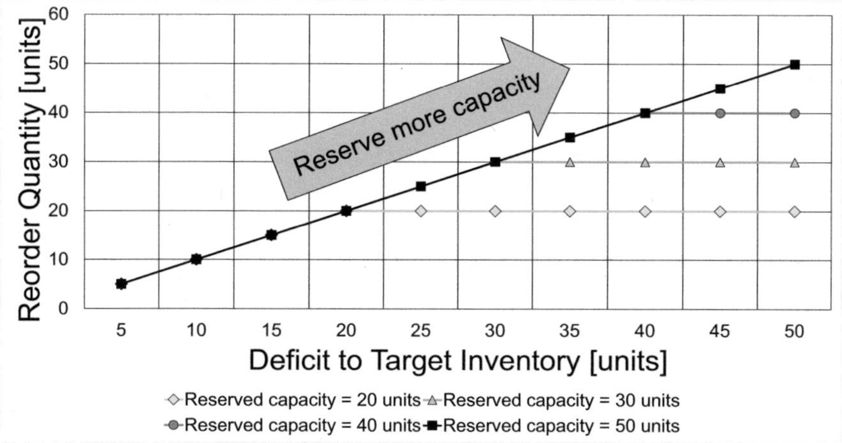

Figure 6.1: Relation between kanban system and heijunka system

In order to ensure the functionality of the system, some conditions must be satisfied. First, the expected value of the reserved capacity in the leveling horizon $E(C)$ must be greater than the expected demand. If $E(C) = E(D)$ or even $E(C) < E(D)$, the expected deficit tends to infinity (see section 4.3.1). The capacity which is reserved in excess of the required capacity (or expected demand) $E(D)$ is called the catch-up capacity C_{CUC}. Therefore, the reserved capacity is the sum of required capacity and catch-up capacity. Furthermore, this catch-up capacity must be greater than zero. Mathematically, this can be expressed as

$$E(C) = E(D) + C_{CUC} \tag{6.1}$$

$$C_{CUC} > 0 \tag{6.2}$$

In the above equations, C_{CUC} denotes for the catch-up capacity in absolute units. The catch-up capacity can also be expressed as a percentage of the reserved capacity. We define the relative catch-up capacity CUC as

$$CUC = \frac{C_{CUC}}{E(C)} \tag{6.3}$$

As stated in section 4.3.1, the behavior of a heijunka leveled kanban system can be described by a G|G|1 queuing system in discrete time. In a queuing system, the utilization is given by $\rho = E(D)/E(C)$. Substituting (6.1) into (6.4) and considering the queuing system's utilization yields the relation between the catch-up capacity and the utilization.

$$CUC = 1 - \rho \tag{6.4}$$

We can now calculate the required buffer inventory as a function of the catch-up capacity (see Figure 6.2).

If we reserve more capacity, we also need more transport capacity to be able to transport the goods that are being ordered. Therefore, the required transport capacity also increases with the catch-up capacity (see Figure 6.2).

In Figure 6.2, both buffer inventory and transport capacity are plotted as physical units which increase or decrease exponentially with increasing catch-up capacity. The costs are linear functions of these physical units. Their slopes correspond to the unit cost rates. There must be a minimum in between the two extremes of an all-inventory or all-capacity buffer. At this minimum, the increase in transport costs due to an increase in catch-up capacity is equal to the decrease of inventory costs. If we change the catch-up capacity starting from the optimum, either the cost of inventory or the cost of transport capacity would increase.

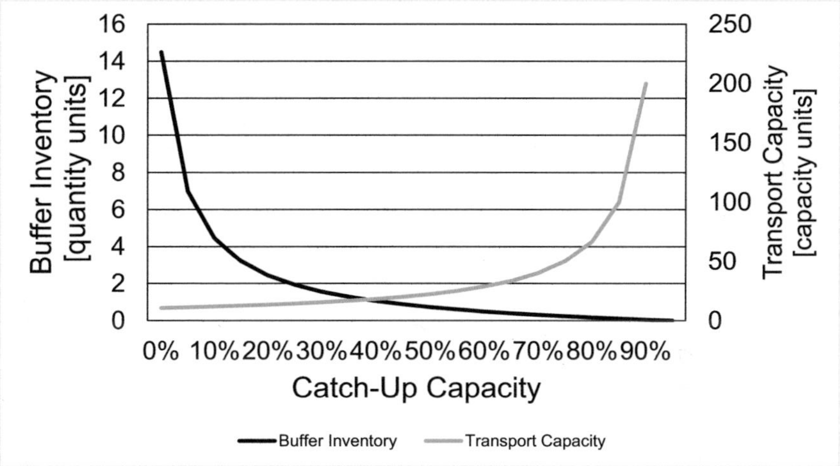

Figure 6.2: Required buffer inventory and transport capacity as a function of catch-up capacity

In the subsequent section, we evaluate the trade-off that has been described above. We develop a mathematical expression of the costs and marginal costs that are caused by both inventory and transport capacity. Based on the marginal costs, we calculate the cost-minimal buffer allocation.

6.2 Modeling the System Operating Costs

The system's total cost of operation K_{TO} is the sum of both inventory costs K_I and costs for transport capacity K_{TR}. To determine an optimum buffer allocation, we need to find mathematical expressions for both terms as a function of the planned utilization ρ or catch-up capacity.

$$K_{TO} = K_{TR} + K_I \tag{6.5}$$

Table 6.1 summarizes all the parameters and their units we will make use of to build the cost model. Both K_{TR} and K_I refer to the costs incurred in the leveling period. They are expressed in terms of money units (MU). In our model, we assume that transports are performed according to the milk run concept. That is, one tour consists of a certain set of suppliers. Over the leveling horizon, the suppliers are not altered and the tour is repeated N_T times. The costs of one tour consist of a fixed and variable component. The fixed costs are expressed in terms of money units per tour, whereas the variable costs are expressed as money units per capacity unit (CU). Moreover, we consider different vehicle sizes. The smallest possible size is $C_{v,min}$, given in capacity units per tour. The biggest possible size is $C_{v,max}$, also given in capacity units per tour.

The inventory model aims at quantifying the total costs of inventory in the leveling period. The inventory holding costs are caused by the capital which is tied due to holding inventory. The costs are quantified as a percentage of the part price for the leveling horizon, and expressed in terms of money units per time unit (TU).

Table 6.1: Overview of parameters and variables employed to model the system's total costs of operation

Subsystem	Symbol	Description	Unit
Transport	K_{TR}	Transport costs in the leveling horizon	MU
	k_f	Fixed costs of transportation of one tour	$MU/Tour$
	k_v	Variable costs of transportation of one tour	MU/CU
	$C_{v,min}$	Smallest possible vehicle capacity	$CU/Tour$
	$C_{v,max}$	Biggest possible vehicle capacity	$CU/Tour$
	C_{Tour}	Capacity which is required for one tour	$CU/Tour$
	TC	Total required capacity in the leveling horizon	CU
	$\widetilde{TC}$	Total chargeable capacity in the leveling horizon	CU
	N_T	Number of tours in the leveling horizon	$Tours$
	$N_{T,max}$	Maximum number of tours in the leveling horizon	$Tours$
Inventory	K_I	Inventory costs in the leveling horizon	MU
	f	Variability parameter	[ ]
	v_a	Coefficient of variation of the inter-arrival time or capacity	[ ]
	v_b	Coefficient of variation of the service time or demand	[ ]
	P	Part price	MU
	i	Inventory holding cost rate in the leveling horizon	$\frac{\%}{TU \cdot Part}$
	k_h	Inventory holding costs of one unit of inventory in the leveling horizon	$\frac{MU}{TU \cdot Part}$
	N_P	Number of different parts that are served by the tour	$Parts$

In the following section we first model the costs of transportation. In the subsequent section, we model the costs of inventory.

6.2.1 Modeling the Costs of Transportation

The cost of transportation consist of various factors. A large fraction of the cost of transportation amounts to the driver's wages. This fraction is more or less fixed – since the drivers are usually paid by hour, it only depends on the duration of the transport. In European countries, drivers are only allowed to drive a maximum of 8 hours per day. Therefore, the assumption of a fixed duration is valid. These fixed costs do not depend on the size of the shipment which has to be transported.

Another fraction of the fixed costs of transportation is the depreciation of the vehicle, i.e. the diminishing value which is caused by wear and tear. This fraction increases with increasing vehicle size: a bigger vehicle is usually more expensive and hence causes more depreciation if the lifetime stays the same. Another factor that increases with vehicle size is the fuel which is consumed by the vehicle. A bigger vehicle is heavier and hence consumes more fuel.

Daganzo (2005) takes these aspects into consideration and proposes to model the costs of one tour for a given distance by a fixed and a variable cost component. As long as the shipment requires a transport capacity which is smaller than the capacity of the smallest vehicle, the cost is equal to the fixed component k_f. That is, one tour with a vehicle capacity $C_{v,min}$ and a fixed duration costs k_f money units for a given distance.

If the shipment requires an amount of capacity which exceeds the smallest vehicle's capacity, we need to choose a vehicle with a higher capacity. This causes variable costs. The costs of the transport increase linearly with the amount by which the vehicle's capacity $C_{v,min}$ is exceeded by the capacity which is required by the goods which need to be transported during the tour C_{Tour}. The slope of the linear increase is given by the variable cost per capacity unit k_v. Mathematically, we can express this as follows.

$$K_{Tour} = \begin{cases} k_f & for\ 0 \leq C_{Tour} \leq C_{v,min} \\ k_f + k_v \cdot (C_{Tour} - C_{v,min}) & else \end{cases} \tag{6.6}$$

Since truck sizes are discrete in reality, the graph of the costs is actually a step function. For reasons of simplicity we approximate this step function by a linear slope (see Figure 6.3). Later in this section, this will enable us to take the first derivative more easily.

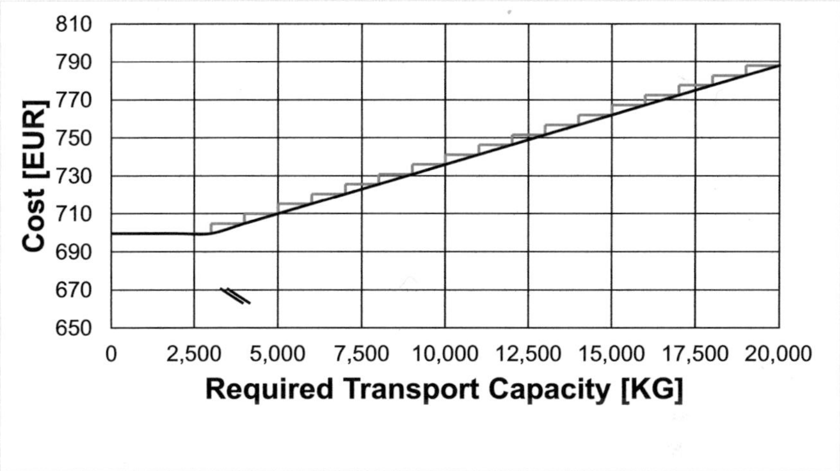

Figure 6.3: Costs of transportation as a function of shipment size.

Typically, a leveling horizon does not only consist of one tour. Therefore we now consider how we can break the total transport capacity which is required in the leveling horizon down to tours in order to be able to quantify K_{TR}. As a first step we need to differentiate between the required transport capacity in the leveling horizon TC and the chargeable transport capacity in the leveling horizon $\widetilde{TC}$.

The required capacity is given by the product of the number of units which are required during the leveling horizon and their unit capacity requirement. Since we want to ensure a catch-up capacity, we need to reserve more capacity that is actually required. Therefore, we define the chargeable transport capacity as the sum of required transport capacity and catch-up capacity (cf. section 6.1). Mathematically, we can express this as

$$\widetilde{TC} = \frac{TC}{1 - CUC} = \frac{TC}{\rho} \tag{6.7}$$

As the next step, we now need to consider how to break the chargeable transport capacity down to tours, i.e. determine the number of transport N_T. It is a function of the vehicle size C_v we choose. There are many different possible (C_v, N_T) combinations which provide the required transport capacity. For the sake of simplicity, we consider two simple strategies in following (cf. section 4.4.3.1).

One possibility is to aim for the highest possible delivery frequency $N_{T,max}$. That is, we assume to use the smallest possible vehicle size $C_{v,min}$ and calculate the number of transports necessary to meet the requirements regarding transport capacity (Strategy A). As long as the chargeable capacity $\widetilde{TC}$ is still below $N_{T,max} \cdot C_{v,min}$, we can use the smallest vehicle and increase the number of tours N_T. If the maximum number of tours $N_{T,max}$ is reached, we need to choose a vehicle with a higher capacity.

The second Strategy is to aim for the lowest possible delivery frequency. That is, we use the biggest possible vehicle $C_{v,max}$ and calculate the resulting number of transports (Strategy B). Whereas the first Strategy primarily aims at minimizing inventory costs, the second one aims at minimizing transport costs. Following these two strategies, the number of tours per month can be calculated by

$$N_T = \begin{cases} min\left(\frac{\widetilde{TC}}{C_{v,min}}, N_{T,max}\right) & for\ Strategy\ A \\ \frac{TC}{C_{v,max}} & for\ Strategy\ B \end{cases} \tag{6.8}$$

Given the number of transports per month N_T, we can calculate the total costs of transportation in the leveling horizon. It is given by the product of the costs of one transport K_{TR} and the number of tours in the leveling horizon N_T.

$$K_{TR} = K_{Tour} \cdot N_T \tag{6.9}$$

In case of Strategy A, we need to consider that the cost function consists of two sections. For a high utilization or low catch-up capacity, the total costs depend on the number of tours and the fixed costs. In this range, using the smallest possible vehicle $C_{v,min}$ provides a sufficient amount of capacity and therefore there are no variable costs. If rho decreases, we reach the maximum number of tours $N_{T,max}$ at some point. When this point is reached, we are now longer allowed to increase the number of tours. This is why we need to choose a vehicle with a high payload to provide the transport capacity. As given by equation (6.6), this causes variable costs.

Since the chargeable weight $\widetilde{TC}$ is a function of ρ, we can calculate the threshold ρ_T at which $N_{T,max}$ is reached. It is given by the point at which $\widetilde{TC} = C_{v,min} \cdot N_{T,max}$. Solving for ρ yields

$$\rho_T = \frac{TC}{C_{v,min} \cdot N_{T,max}} \tag{6.10}$$

We note that if $TC > C_{v,min} \cdot N_{T,max}$, ρ_T is greater than one. That is, even with a catch-up capacity of zero, we need to choose a vehicle with a capacity $C_v > C_{v,min}$ and the cost function only consists of only section. If $TC < C_{v,min} \cdot N_{T,max}$, we need to distinguish the two sections. If $\rho \rightarrow 1$, i.e. $CUC \rightarrow 0$, the maximum number of tours $N_{T,max}$ is not yet reached. If ρ decreases, we can simply increase the number of tours N_T. At $\rho = \rho_T$, we have reached the maximum number of tours N_T. If ρ decreases further, we need to choose a vehicle with a capacity $C_v > C_{v,min}$.

We can now express the costs of transportation K_{TR} as a function of ρ. Therefore, we substitute expressions (6.7) and (6.8) into (6.6) for Strategy A, which yields the following:

$$K_{TR}(\rho) = \begin{cases} N_{T,max} \cdot k_f - k_v \cdot C_{v,min} + k_v \cdot \frac{TC}{\rho} & for\ 0 \leq \rho < \rho_T \\ \frac{TC}{\rho \cdot C_{v,min}} \cdot k_f & for\ \rho_T \leq \rho \leq 1 \end{cases} \tag{6.11}$$

The costs of Strategy B behave similar to the first section of Strategy A. The only difference is that, since we use a vehicle of payload $C_{v,max}$, the fixed costs are different. We calculate the fixed costs of Strategy B k_f' and the costs of transportation $K_{TR}(\rho)$ as follows

$$k_f' = k_f + k_v \cdot \left(C_{v,max} - C_{v,min}\right) \tag{6.12}$$

$$K_{TR}(\rho) = \frac{TC}{\rho \cdot C_{v,min}} \cdot k_f' \quad for\ \rho \in [0,1] \tag{6.13}$$

Figure 6.4 shows the graph of K_{TR} as a function of the chargeable capacity $\widetilde{TC}$ for both strategies. For the quantitative example we assumed a leveling period of one month. The smallest possible vehicle payload is $C_{v,min} = 3$ tons and the maximum number of tours in the leveling period is $N_{max} = 20$ tours per month. The cost rates are taken from Wilken (2017).

We note that Strategy B is characterized by a single slope whilst the slope of Strategy A changes at 60 tons. The reason is that until the monthly weight of 60 tons is reached, Strategy A tells us to use the smallest possible vehicle with a payload of 3 tons and change the number of tours per month. Since the proportion of fixed costs is higher than the proportion of variable costs, the slope is higher in this range. As soon as the threshold of 60 tons per month is surpassed, the number of transports per month stays at $N_{T,max} = 20$ and we opt for a vehicle with a higher payload (see equation (6.8)).

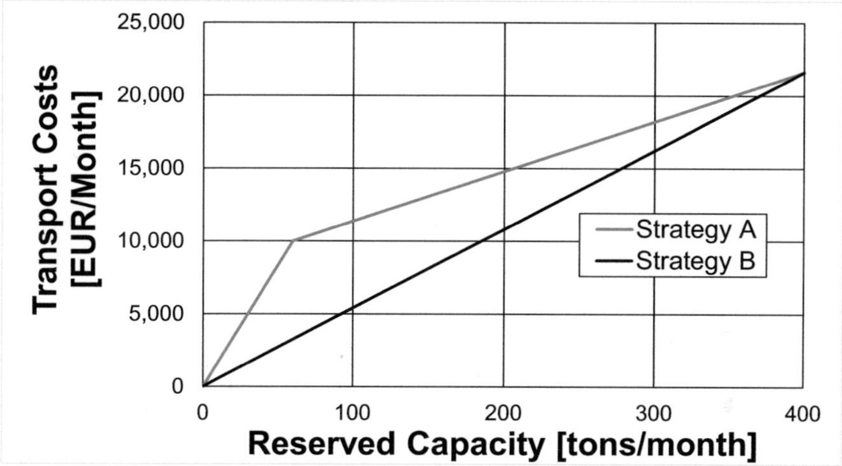

Figure 6.5: Transport costs per month as a function of the chargeable capacity $\widetilde{TC}$ for strategies A and B

6.2.2 Modeling the Costs of Inventory

As stated in section 4.3.1, a heijunka leveled inventory system behaves in line with a G|G|1 queuing system. According to the analogy, the deficit to the target inventory after satisfying the period demand is equivalent to the number of customers in the system. In contrast to the preceding sections, in which we used a discrete time G|G|1 model for our calculations, we employ a continuous time G|G|1 model in this section. Due to the continuization, we are able to describe the average number of customers in the queuing system as a continuous function of the utilization. Later, this enables us to find a more simple expression for the derivative of the cost function.

There are different approximation methods to calculate the average number of customers in the system for a continuous time G|G|1 queuing system (cf. Arnold and Furmans, 2009). One way to calculate the continuous time G|G|1 performance figures is the approximation method of Gudehus (1976). It uses the first two moments of the arrival and service process to estimate performance indicators of the queuing system. The model's input parameters are the arrival stream's expected interarrival time $E(t_a)$, its coefficient of variation v_a, the expected service time $E(t_b)$, and its coefficient of variation v_b. Moreover, we can calculate the system's average utilization $\rho = E(t_b)/E(t_a)$.

As described in section 4.3.1, the interarrival time corresponds to the capacity and the service time corresponds to the demand. We must note, however, that the average number of customers N_S does not correspond to the average inventory but to the average deficit (cf. section 4.3.1). To calculate the average inventory, we need to consider the number of kanbans of the system. Since we cannot calculate quantiles for continuous G|G|1 queuing systems, this is not possible. Each time the deficit is greater than the number of kanbans, the inventory is zero. For a sufficiently high number of kanbans, i.e. a sufficiently high service level, the difference between average deficit and average inventory decreases. Therefore, for the following calculations, we assume a sufficiently high service level to approximate the average inventory by the average deficit.

Following Gudehus (1976), the average number of customers in a continuous time G|G|1 queuing system N_S can be expressed as a function of ρ, v_a^2 and v_b^2. Under consideration of (6.4), we can also express N_S as a function of the catch-up capacity CUC.

$$
\begin{aligned}
N_S &= \frac{\rho}{1-\rho}\left(1-\rho\left(1-\frac{v_a^2+v_b^2}{2}\right)\right) \\
N_S &= \frac{(1-CUC)}{CUC}\left(1-(1-CUC)\left(1-\frac{v_a^2+v_b^2}{2}\right)\right)
\end{aligned}
\tag{6.14}
$$

To calculate the inventory costs in the leveling horizon, we need to introduce the inventory holding costs per unit k_h. It is given as the product of the price per unit P and inventory holding cost rate for the leveling horizon i.

$$k_h = P \cdot i \tag{6.15}$$

To calculate the inventory costs in the leveling horizon K_I, we multiply N_S, i.e. the average inventory, with the inventory holding cost rate per unit k_h. Moreover, we can further simplify the term by introducing the variability parameter $f = 1 - (v_a^2 + v_b^2)/2$.

$$\begin{aligned} K_I &= N_S \cdot k_h \\ &= \frac{\rho}{1-\rho}(1 - \rho \cdot f) \end{aligned} \tag{6.16}$$

Figure 6.6 displays N_S as a function of the utilization ρ for different variability parameters f. We note that the higher the absolute value of f, the higher the increase of queue length with utilization.

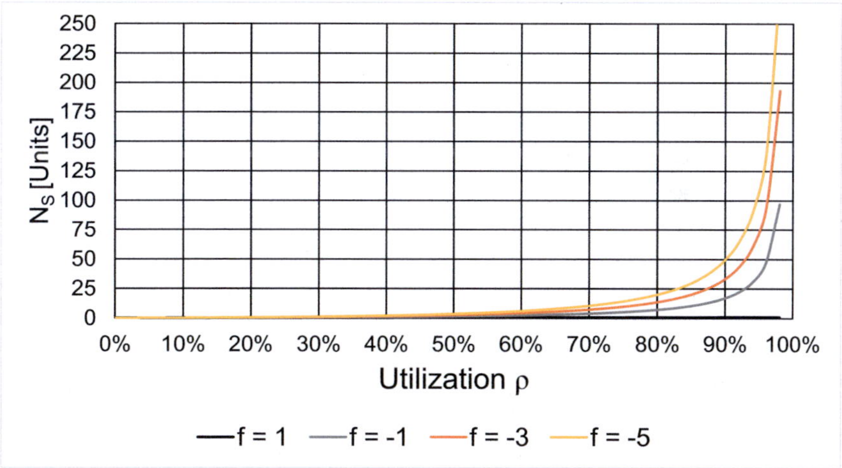

Figure 6.6: Queue length as a function of the utilization for different variability parameters f

In case of multiple products, the variability parameter f and the holding costs k_h are different for each product i. To calculate the inventory holding costs for N_P products, we need sum up the inventory holding costs of all different products.

$$K_I = \sum_{i=1}^{N_P} \frac{\rho}{1-\rho}(1 - \rho \cdot f_i) \cdot k_{h,i} \tag{6.17}$$

If we assume the same holding costs per unit for all products that is $k_{h,i} = k_h$ for all i, and by introducing $\bar{f}$ as the arithmetic mean of the variability parameter of all products, we can simplify the expression as follows:

$$K_I = N_P \cdot \frac{\rho}{1-\rho}\left(1 - \rho \cdot \bar{f}\right) \cdot k_h \tag{6.18}$$

6.3 Determining the Pareto-Efficient Buffer Allocation

In the preceding sections, we derived mathematical expressions for the cost of transportation and the costs of inventory. We can now combine both to find an expression for the system's total cost of operations for both strategies A and B.

The total costs of Strategy A can be derived by combining equations (6.16) and (6.11) with (6.5). As stated in section 6.2.1, the transport curve for Strategy A consists of two sections. If the chargeable capacity $\widetilde{TC}$ is smaller than $C_{v,min} \cdot N_{T,max}$, the number of tours per month are varied as response to a change in ρ. If the chargeable capacity exceeds $C_{v,min} \cdot N_{T,max}$, we need to use a vehicle with a higher capacity. This can be expressed as follows

$$K_{TO}(\rho) = \begin{cases} N_{T,max} \cdot k_f - k_v \cdot C_{v,min} + k_v \cdot \dfrac{TC}{\rho} \\ + \dfrac{\rho}{1-\rho}\left(1 - \rho \cdot \bar{f}\right) \cdot N_P \cdot k_h & for\ 0 \leq \rho < \rho_T \\ \dfrac{TC}{\rho \cdot C_{v,min}} \cdot k_f + \dfrac{\rho}{1-\rho}\left(1 - \rho \cdot \bar{f}\right) \cdot N_P \cdot k_h & for\ \rho_T \leq \rho \leq 1 \end{cases} \quad (6.19)$$

Likewise, we combine (6.16), (6.13) and (6.5) for Strategy B and obtain

$$K_{TO}(\rho) = \frac{1}{\rho} \cdot \frac{TC}{C_{v,max}} \cdot k_f' + \frac{\rho}{1-\rho}\left(1 - \rho \cdot \bar{f}\right) \cdot N_P \cdot k_h \qquad for\ \rho \in [0,1] \quad (6.20)$$

All parameters $TC, C_{v,min}, C_{v,max}, f, k_v$ and k_h are greater than zero, ρ is only defined between 0 and 1 and $\bar{f} \leq 1$. Therefore, we note two properties of the total cost functions $K_{TO}(\rho)$.

$$\lim_{\rho \to 0} K_T(\rho) = \infty \quad (6.21)$$

$$\lim_{\rho \to 1} K_T(\rho) = \infty \quad (6.22)$$

Since it is only defined in the interval between 0 and 1, the total cost function is convex. Therefore a local minimum exists between 0 and 1. Moreover, since the function tends to infinity, we do not need to check the boundaries $\rho = 0$ and $\rho = 1$ for a minimum.

Figure 6.7 displays the total costs as a function of the catch-up capacity for different input parameters. We note that at the costs tend to infinity at the left and right boundaries. Moreover, we note that the minimum is the located inbetween.

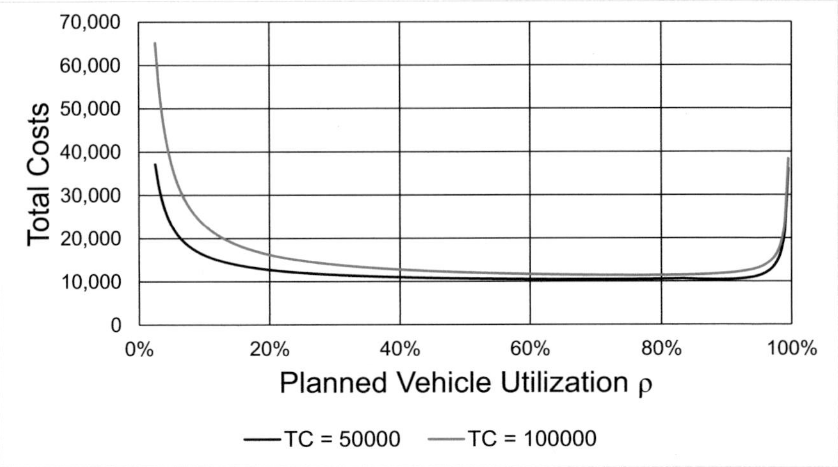

Figure 6.7: Total costs as a function of catch-up capacity for different input parameters

The necessary condition for a minimum is that the slope of the cost curve, given by the marginal cost function $K_{TO}'(\rho)$, takes a value of zero. Therefore, to obtain the utilization ρ_{min} which minimizes the total costs, we take the first derivative with respect to ρ.

In case of Strategy A, we need to take into account the distinction of cases and calculate the derivative of each of the two sections.

$$K_{TO}'(\rho) = \begin{cases} \dfrac{-TC \cdot k_v}{\rho^2} + \dfrac{(\bar{f}\rho^2 - 2\rho\bar{f} + 1)N_P k_h}{(1-\rho)^2} & for\ 0 \leq \rho < \rho_T \\ \dfrac{-TC \cdot k_f}{C_{v,min} \cdot \rho^2} + \dfrac{(\bar{f}\rho^2 - 2\rho\bar{f} + 1)N_P k_h}{(1-\rho)^2} & for\ \rho_T \leq \rho \leq 1 \end{cases} \quad (6.23)$$

In case of Strategy B, the function of the total costs of operations only consists of one section. The derivative with respect to ρ is given by

$$K'_{TO}(\rho) = \frac{(-1)}{\rho^2} \cdot \frac{TC \cdot k'_f}{C_{v,max}} + \frac{(\bar{f}\rho^2 - 2\rho\bar{f} + 1)N_P \cdot k_h}{(1-\rho)^2} \quad for\ \rho \in [0,1] \tag{6.24}$$

The necessary condition for a local minimum is $K'_T(\rho) = 0$. Substituting and re-arranging of (6.23) yields.

$$K'_{TO}(\rho) = 0$$
$$= \begin{cases} \bar{f}\rho^4 - 2\bar{f}\rho^3 + \rho^2\left(1 - \frac{TC \cdot k_v}{N_P \cdot k_h}\right) + \frac{2\rho \cdot TC \cdot k_v - TC \cdot k_v}{N_P \cdot k_h} & for\ 0 \le \rho < \rho_T \\ \bar{f}\rho^4 - 2\bar{f}\rho^3 + \rho^2\left(1 - \frac{TC \cdot k_f}{C_{v,min} N_P k_h}\right) + \frac{2\rho \cdot TC \cdot k_f - TC \cdot k_f}{C_{v,min} \cdot N_P \cdot k_h} & for\ \rho_T \le \rho \le 1 \end{cases} \tag{6.25}$$

Likewise, we substitute $K'_T(\rho) = 0$ and re-arrange (6.24) in case of Strategy B

$$K'_{TO}(\rho) = 0$$
$$= f\rho^4 - 2f\rho^3 + \rho^2 \cdot \left(1 - \frac{TC \cdot k'_f}{C_{v,min} \cdot N_P \cdot k_h}\right) + \frac{2\rho \cdot TC \cdot k'_f - TC \cdot k'_f}{C_{v,min} \cdot N_P \cdot k_h} \quad for\ \rho \in [0,1] \tag{6.26}$$

The minimum of the total cost function K_{TO} is given by solving equations (6.25) and (6.26) for ρ. However, we note that both equations are fourth degree polynomials. They can be solved analytically for instance by the method given in Shmakov (2011), but the formulas are quite long and unwieldy. This complicates the study of the system behavior. This is why we will present two simple methods to solve the equation in the subsequent sections. First, we numerically compute the location of the optimum as a function of the input parameters. After that, we present how to simplify the fourth degree polynomial and linearize the equation by means of a Taylor series to calculate an analytical optimum.

6.3.1 Numerical Calculation

In this section we calculate a numerical solution to the fourth-degree polynomial which is given by equations (6.31) and (6.32). To find a solution, we enumerate over a variety of practically relevant factor cost ratios and determine the zeroes of the marginal cost functions.

Both expressions (6.31) and (6.32) contain many parameters which make the expression long and unwieldy. To simplify the expressions, we summarize certain parameters by one symbol and introduce a new kind of parameter, the factor costs of both inventory and transportation. All factor costs are denoted by the Greek letter κ and differentiated by their respective index. The factor costs denote the rate by which the total costs change for a change of ρ. For the costs of transportation of Strategy A we introduce $\kappa_{T,f,A}$and $\kappa_{T,v}$ for the two sections we need to differentiate.

$$\kappa_{TR,f,A} = \frac{TC \cdot k_f}{C_{v,min}} \tag{6.27}$$

$$\kappa_{TR,v} = TC \cdot k_v \tag{6.28}$$

The cost function in case of Strategy B consists of only one section, therefore we only introduce the single factor cost rate $\kappa_{TR,f,B}$.

$$\kappa_{TR,f,B} = \frac{TC \cdot k_f'}{C_{v,max}} \tag{6.29}$$

The inventory costs behave the same for both Strategies. Their factor cost rate is denoted by κ_I and given by

$$\kappa_I = N_P \cdot k_h \tag{6.30}$$

Substituting these into expression (6.25) and further dividing by $\bar{f}$ for Strategy A yields

$$K'_{TO}(\rho) = 0 = \begin{cases} \rho^4 - 2\rho^3 + \frac{\rho^2}{\bar{f}}\left(1 - \frac{\kappa_{TR,v}}{\kappa_I}\right) + \frac{2\rho \cdot \kappa_{TR,v}}{\kappa_I \cdot \bar{f}} - \frac{\kappa_{TR,v}}{\kappa_I \cdot \bar{f}} & for\ 0 \leq \rho < \rho_T \\ \rho^4 - 2\rho^3 + \frac{\rho^2}{\bar{f}}\left(1 - \frac{\kappa_{TR,f}}{\kappa_I}\right) + \frac{2\rho \cdot \kappa_{TR,f}}{\kappa_I \cdot \bar{f}} - \frac{\kappa_{TR,f}}{\kappa_I \cdot \bar{f}} & for\ \rho_T \leq \rho \leq 1 \end{cases} \quad (6.31)$$

Substituting likewise for Strategy B yields

$$K'_{TO}(\rho) = 0 = \rho^4 - 2\rho^3 + \frac{\rho^2}{\bar{f}}\left(1 - \frac{\kappa_{TR,f}}{\kappa_I}\right) + \frac{2\rho \cdot \kappa_{TR,f}}{\kappa_I \cdot \bar{f}} - \frac{\kappa_{TR,f}}{\kappa_I \cdot \bar{f}} \quad for\ \rho \in [0,1] \quad (6.32)$$

We note that for both equations (6.31) and (6.32), the location of the optimum ρ_{min} only depends on the ratio of the factor costs of transport cost to factor costs of inventory and the variability parameter $\bar{f}$. Moreover we note that all equations have the same structure. The only difference are the constants. This why we treat all factor costs of transport as the same for our enumeration. They represent different points on one curve. We need to keep in mind that for Strategy A, there might exist two local minima, one for each section. Therefore, we need to look up the minimum catch-up capacity for both $\kappa_{T,f,A}$ and $\kappa_{T,v}$ to find the two local optima.

Figure 6.8 shows the optimum catchup-capacity as a function of the relation κ_T/κ_I for different values of the variability indicator f. We note that in the interval between 0 and 2000, the slope of the curves is highest. With increasing κ_T/κ_I ratio, the slopes of the curves eventually decrease. We further note that for κ_T/κ_I approaching infinity, the curves converge asymptotically to a catch-up capacity of 0.

We also note that the influence of the variability parameter f decreases with increasing κ_T/κ_I. For κ_T/κ_I tending to zero, the difference between the curves is greatest. At about κ_T/κ_I, all curves have an optimum catch-up capacity below zero.

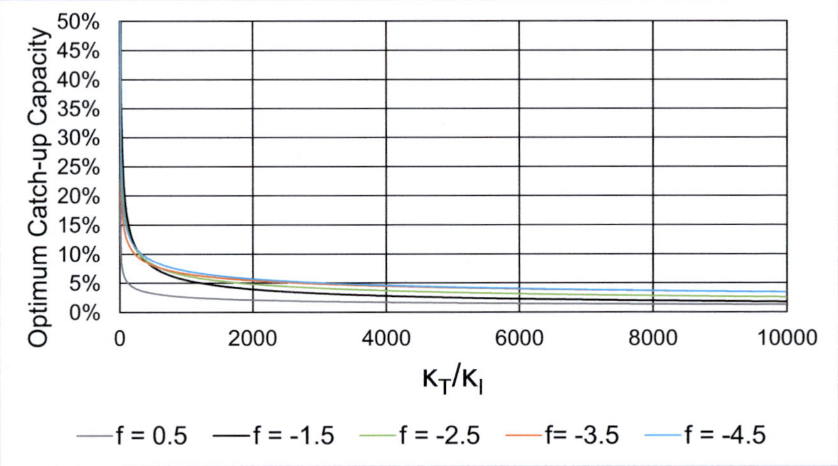

Figure 6.8: Optimum catch-p capacity as a function of factor cost ratio κ_T/κ_I

Figure 6.8 can be used to look up the optimum catch-capacity for given input parameters. In Table 6.2, we present two examples for a different monthly throughput TC which illustrate this process. We assume a length of the leveling horizon of one month and a maximum of $N_{T,max} = 20$ tours. For both examples we use planning Strategy A. The minimum vehicle capacity we consider is $C_{v,min} = 3{,}000\ KG$. Therefore, we are able to use a truck with a payload of $C_{v,min}$ up to a total monthly capacity of $TC = 60\ tons$. If TC

exceeds 60 tons, we need to use a larger truck because the maximum number of tours per month $N_{T,max}$ is reached. Moreover, we consider a variability parameter of $f = -1.5$.

Table 6.2: Numerical example to illustrate the usage of Figure 6.8

Subsystem	Parameter	Example 1		Example 2
Transport	$TC\ [KG]$	50,000		200,000
	$N_{T,max}[Tours]$	20		
	$C_{v,min}[KG]$	3,000		
	$k_f\ [EUR/Tour]$	500		
	$k_v\ [EUR/KG]$	0.014		
	$\kappa_T[EUR]$	8,333	700	2,800
Inventory	$Unit\ Price[EUR/Part]$	50		
	$Number\ of\ parts$	100		
	$Holding\ cost\ rate\ [\%\ per\ year]$	10		
	$\kappa_I\ [EUR]$[1]	3.99		
Results	κ_T/κ_I	2,088	700	702
	CUC_{opt}	4%	6%	6%

The first step is to calculate the factor costs of transportation κ_T. In case of Example 1, the monthly throughput TC is below sixty tons. Therefore we need to consider both sections of the cost function to determine the optimum. For Example 2, TC exceeds 60 tons. Therefore, we only need to consider the variable costs k_v to determine the optimum. Given κ_T, we need to determine κ_I. This is the total inventory costs in the leveling horizons, i.e. the inventory costs of all parts which are procured by the transport. For our example we assume 100 different parts which are procured and all have a unit

[1] We calculated k_h as follows. The annual interest rate is 10%. Therefore the monthly interest rate is $1.1^{\frac{1}{12}} - 1 = 0.00797, 100\ Units \cdot 0.797 \frac{\%}{month} \cdot 50 \frac{EUR}{Unit} = 3.99 \frac{EUR}{month}$.

price of EUR 50. The annual inventory cost rate rate[2] is 10%. Given these values, the factor costs κ_I and hence also the ratio κ_T/κ_I can be determined. The optimum catch-up capacity can now be read from Figure 6.8 for all factor cost ratios.

In case of example 1, the slope of the cost function changes at $\rho_T = 83.3\%$ or $CUC = 16.7\%$.That is, in the range between $CUC = 100\%$ and $CUC = 16.7\%$, the factor costs of transportation depend on the variable costs k_v. The optimum that is yielded by considering k_v is, however, at $CUC = 6\%$. At this point, the cost of transportation do not depend on k_v but only on k_f. Therefore, this point cannot be the optimum. The true optimum is given by $CUC = 4\%$.

6.3.2 Taylor Series Approximation

In addition to the numerical solution of the preceding chapter, we present an analytical solution in this chapter. As (6.32) is a fourth order polynomial, the solution is a rather long and unwieldy term. To reduce the order of the polynomial, we linearize the equation around the point of expansion ρ_0 by developing a Taylor series.

In general, a linear Taylor series of a function f can be used to approximate the function by calculating the function's value in the point of expansion $f(\rho_0)$ and the function's slope in the point of expansion $f'(\rho_0)$. That is, we assume that the function is linear in the point of expansion and a close range around this point of expansion. Mathematically, this can be expressed as follows:

$$f(\rho) = f(\rho_0) + f'(\rho_0) \cdot (\rho - \rho_0) \tag{6.33}$$

In our case, we linearize $K'(\rho)$ to be able to determine the zeros. To develop our Taylor series, we need to calculate $K'(\rho_0)$ and the slope $K''(\rho_0)$in the operating point ρ_0.

[2] According to Weber (2012) any annual inventory interest rate between 4 and 20% can be justified easily on the basis of different economic models.

$$K_{TO}{}'(\rho) = K'_{TO}(\rho_0) + K''_{TO}(\rho_0) \cdot (\rho - \rho_0) \tag{6.34}$$

Our goal is to find the zeroes of the marginal cost function. This why we substitute $K^{'}(\rho) = 0$ into (6.34), and re-arrange the equation for ρ.

$$\begin{aligned} 0 &= K'_{TO}(\rho_0) + K''_{TO}(\rho_0) \cdot (\rho - \rho_0) \\ \rho &= \rho_0 - \frac{K'_{TO}(\rho_0)}{K''_{TO}(\rho_0)} \end{aligned} \tag{6.35}$$

The expression $K'_{TO}(\rho_0)$ in equation (6.35) can be obtained by substituting ρ_0 into (6.23). The denominator $K''_{TO}(\rho_0)$ can be obtained by taking the derivative of (6.23) with respect to ρ. For Strategy A this yields

$$K''_{TO}(\rho) = \begin{cases} \dfrac{2 \cdot TC \cdot k_v}{\rho^3} + \dfrac{2 \cdot ((f-1)\rho + 1 - f) \cdot k_h \cdot N_P}{(1-\rho)^4} & for\ 0 \leq \rho < \rho_T \\ \dfrac{2 \cdot TC \cdot k_f}{\rho^3 \cdot C_{v,min}} + \dfrac{2 \cdot ((f-1)\rho + 1 - f) \cdot k_h \cdot N_P}{(1-\rho)^4} & for\ \rho_T \leq \rho \leq 1 \end{cases} \tag{6.36}$$

Likewise, for Strategy B, we need to take the derivative of (6.24) with respect to ρ, which yields the following expression.

$$K''_{TO}(\rho) = \frac{2 \cdot TC \cdot k'_f}{\rho^3 \cdot C_{v,max}} + \frac{2 \cdot ((f-1)\rho + 1 - f) \cdot k_h \cdot N_P}{(1-\rho)^4} \tag{6.37}$$

Given expressions (6.36) and (6.37), we can solve equation (6.35) for ρ for both Strategies A and B. In case of Strategy A, ρ can be calculated by

$$\rho = \begin{cases} \rho_0 \cdot \dfrac{E_1 \cdot k_v + E_2 \cdot C_{v,min} \cdot \rho_0^2}{E_3 \cdot k_v + E_2 \cdot C_{v,min} \cdot \rho_0^3} & for\ 0 \leq \rho \leq \rho_T \\ \rho_0 \cdot \dfrac{E_1 \cdot k_f + E_2 \cdot C_{v,min} \cdot \rho_0^2}{E_3 \cdot k_f + E_2 \cdot C_{v,min} \cdot \rho_0^3} & for\ \rho_T \leq \rho \leq 1 \end{cases} \tag{6.38}$$

$$\begin{aligned} &with\ expressions \\ &E_1 = -TC \cdot (1-\rho_0)^2 \\ &E_2 = k_h \cdot N_P \cdot (f\rho_0^2 - 2\rho_0 f + 1) \\ &E_3 = 2 \cdot TC \cdot (1-\rho_0)^4 \end{aligned}$$

Likewise, for Strategy B, ρ is given by

$$\rho = \rho_0 \cdot \frac{E_1 \cdot k_f' + E_2 \cdot C_{v,max} \cdot \rho_0^2}{E_3 \cdot k_f' + E_2 \cdot C_{v,max} \cdot \rho_0^3} \tag{6.39}$$

The accuracy of the linearized solution developed in this section depends to a large degree on the choice of an appropriate point of expansion ρ_0. In general, the point should be set as close to the "best guess solution" as possible. We know from Figure 6.8 of the preceding section that for a wide range of parameter combinations, the optimum is between $CUC = 0$ and $CUC = 10\%$. That is why we suggest to set the point of expansion ρ_0 at some value between 100% and 90%.

6.4 Conclusion

In this chapter we evaluated the efficiency of heijunka leveling as a method of stabilization for systems of transport logistics. We explained the basics of efficiency and defined how the term has to be employed according to Pareto. Moreover, we transferred the concept of Pareto-efficiency to the scope of transport logistics.

We analyzed the kanban system and the heijunka-leveled kanban system and found that both are not two discrete distinct systems but rather different points on a continuum. Both systems are linked by the catch-up capacity: In a heijunka-leveled kanban system, the catch-up capacity delimits the variations that are passed to the upstream stages. In a kanban system, the variations that are passed to the upstream stages is unlimited for a sufficiently high number of kanbans. Therefore, the kanban system is in fact a heijunka system with a certain catch-up capacity. The higher the catch-up capacity, the lower the required buffer inventory but the higher the capacity buffer and less pronounced is the effect of leveling. That is, the more the heijunka system behaves like a kanban system. If the maximum order quantity is equal to the number of kanban cards in the system, both systems behave the same.

Both buffer capacity and buffer inventory are associated with costs. Since both are conflicting objectives, the optimum is characterized by a trade-off. In the minimum of the cost function, a change in catch-up capacity leads to a change of transport costs which is equal to the change in inventory costs, i.e. the slope of the cost function is zero.

To determine this cost-minimal catch-up capacity, we modeled the system operating cost by considering the cost of transportation and the cost of inventory. The cost of transportation were modeled as a linear function consisting of both fixed transport costs and variable transport costs. We differentiated between two different planning strategies and expressed the costs as a function of the required capacity per month and the catch-up capacity. The cost of inventory was calculated by approximating the average required inventory by the average deficit of a continuous-time G|G|1-model and the method of Gudehus (1976). The total system operating costs were obtained by summing up transport costs and inventory costs.

The optimum catch-up capacity was determined by taking the derivative of the function of system operating cost with respect to the catch-up capacity to calculate the marginal cost function. Since the marginal cost function is a fourth-order polynomial, we employed different methods to calculate the zeroes. The problem was enumerated for different input parameters as a function of the ratio of factor costs of inventory and transport. Moreover, we employed a Taylor series to linearize the function and obtain a simpler solution of the equation.

We found that, for a wide range of factor cost ratios, there is a Pareto-optimum between an all-capacity and an all-inventory buffer at a catch-up capacity of about 5%. The location of the optimum depends on the factor cost ratio of transport to inventory. For a factor costs ratio tending to infinity, the optimum catch-up capacity tends to zero. For a factor costs ratio tending to zero, the optimum catch-up capacity tend to infinity. The results show that heijunka leveling, i.e. a limitation of the reorder quantity, is always efficient except for a factor cost ratio tending to zero. The key is to reserve the amount of capacity which corresponds to the Pareto-efficient buffer allocation.

To illustrate the practical relevance of the results which were obtained in this chapter and to check the validity of the mathematical models we presented in this and the preceding sections, we will apply the methods and models to a numerical example in a case study of the German Automotive supplier ZF Friedrichshafen in the following section.

7 Design for Stability by the Example of ZF Friedrichshafen

Kids, you tried your best but you failed miserably.
The lesson is: never try.
(Homer Simpson, TV character)

In the preceding chapter, we created an abstract model of the total costs of a heijunka-leveled transport logistics systems to investigate the trade-off between buffer inventory and buffer capacity. We used this model to determine the Pareto-efficient buffer allocation in terms of the catch-up capacity. As a special case and a numerical example of the abstract model presented in the preceding chapter, we conduct a case study of the German Automotive Supplier ZF Friedrichshafen in this chapter.

In the first section, the sample data provided by ZF Friedrichshafen is introduced by conducting some descriptive analyses. Then we design the system by following the steps that were presented in 4.4. Given the system design, we calculate the system operating costs for a given catch-up capacity. After that, we show how to determine the optimum catch-up capacity by different methods. That is, we use both our exact discrete model and evaluate the approximation accuracy of the continuous model. In the last section, we compare the transport costs of the system we designed with the operating cost of our system if it would be run by an area freight forwarder (see section 3.1.2).

7.1 Introduction to Case Study Data

As the first step of our efficiency analysis, we want to present the case study data provided by ZF to give the reader an understanding of the data. In the first part of this section, we describe the structure of the data. Afterwards, we perform some descriptive analyses.

7.1.1 Structure of Case Study Data

Our case study data encompasses the delivery schedules and part master data of 57 German suppliers of ZF. From these suppliers, a total of 795 parts are procured. The material supply is controlled according to the MRP algorithm (see section 3.2.1). That is, a master production schedule is created and the parts are ordered in advance before the day that they are needed. The temporal offset is equal to the transport lead time. The transports are performed by one area freight forwarder who is responsible for the transports of all the German ZF sites.

The process of ordering material and transporting it to the receiving plant consists of the following steps (c.f. sections 3.1.2 and 3.2.1):

- The **receiving plant** sends a delivery schedule to the supplier. The delivery schedule contains information regarding the part number, its quantity and the date it needs to arrive at the ZF plant.
- Given the required arrival date, the **supplier** calculates a pickup date which ensures that the shipment arrives at ZF on time. Pickup date and information regarding the shipment size (weight, volume) are communicated to the **carrier**.
- The **carrier** picks up the goods at ZF as a part of a pre-leg tour. The goods are brought to a consolidation hub.
- In the main leg tour on the subsequent day, the goods are delivered to ZF.

The structure of the delivery schedule is depicted Table 7.1. It contains a time series of required quantities for each part. The delivery schedule has been exported from the ERP system and contains a schedule for orders with a foresight of one and a half years. However, the further we look into the future, the more inaccurate the schedule gets. Since starting from week 7, monthly quantities are all ordered virtually on one day of the month, we only consider the first six weeks for detailed analysis.

Table 7.1: Structure of a delivery schedule created by the MRP run of the ERP system

Day	Part 1	Part 2	Part 3
1		[quantity]	
2	[quantity]		[quantity]
3		[quantity]	
4	[quantity]		

In addition to the delivery schedule, we were provided part master data. That is, for each part, we have the following information:

- Weight per part [kg]
- Kind of shipment unit (e.g. pallet or plastic box)
- Parts per shipment unit
- Weight of empty shipment unit
- Dimensions of shipment unit (m x m x m)
- Supplier name and address information

Given these data, we have all required information to design and evaluate the system. Since we have ERP information which looks into the future for an interval of about six weeks in combination with the part master data, we are happy to have a data set which we would also use in practice to design a heijunka leveled material supply system.

7.1.2 Descriptive Analyses

After having presented the structure of the data set in the preceding section, we present some descriptive analyses of the data in this section.

Figure 7.1 shows the average order frequencies of all parts in the as-is situation. It was calculated by dividing the number of orders in the sample by the total number of periods. We note that only a small fraction is ordered weekly. Most parts are ordered biweekly, trimonthly or biannual. About 25% percent of the parts is only orders twice a year.

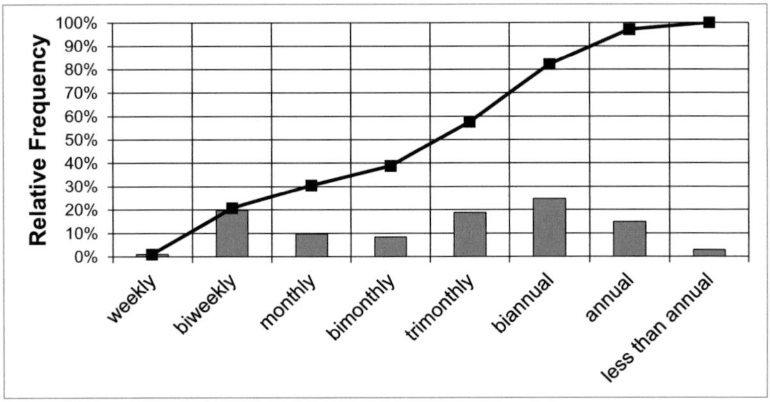

Figure 7.1: Average order frequency in delivery schedule

Figure 7.2 shows a histogram of the average inventory coverage of the minimum order quantity. It was calculated by dividing the minimum order quantity [pieces] by the average daily demand [pcs/day]. We note that almost fifty percent of parts have a MOQ-Coverage of one week. This percentage is higher than the percentage of parts that is ordered weekly. This observation is an indicator that the system might benefit from heijunka leveling. The higher the MOQ that is required by the supplier, the higher the orders we place and the lower their frequency. That is in the status quo we were allowed to order more frequently in the sense of supplier restrictions, but the potential is not leveraged. The reason for this are cost issues. The AFF gives a discount on high shipment sizes. Therefore, it is cheaper to place orders which result in high shipment sizes.

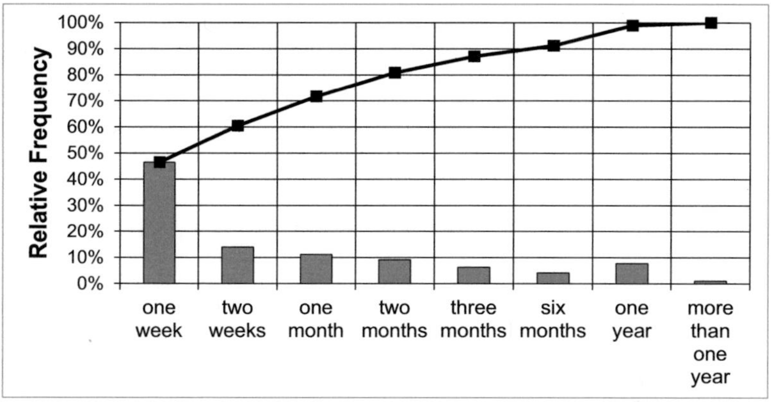

Figure 7.2: Coverage of Minimum Order Quantity

Figure 7.3 shows a histogram of the consumption per individual part, measured in KG/Day. We note that the consumption is unequally distributed. There are a few parts accounting for most of the consumption. The Gini coefficient amounts to G=88.4%. The graph shows that 80% of the parts are consumed at a rate lower than 50 KG/Day.

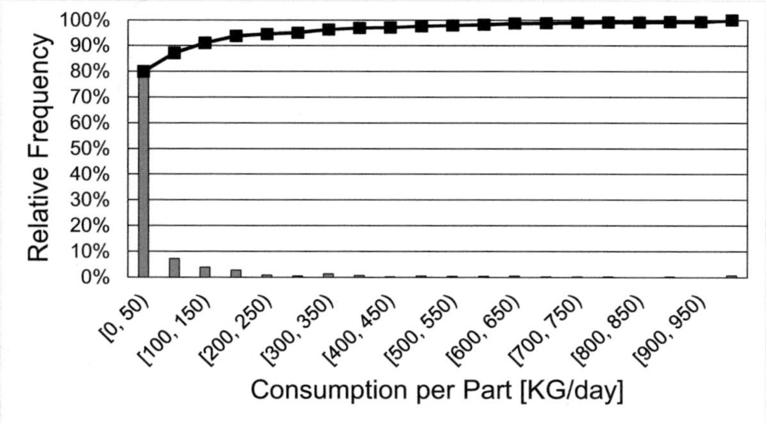

Figure 7.3: Consumption per part [KG/Day]

Figure 7.4 shows the coefficient of variation of the ordered quantities per day. We note that the highest bar is on the right-hand-side of the graph. Most of the parts are only ordered once in the period of observation. This is why their cv^2 is equal to the sample size[1], which is greater than two periods. This observation is consistent with Figure 7.3 which also shows that the product portfolio consists of different high and low runners.

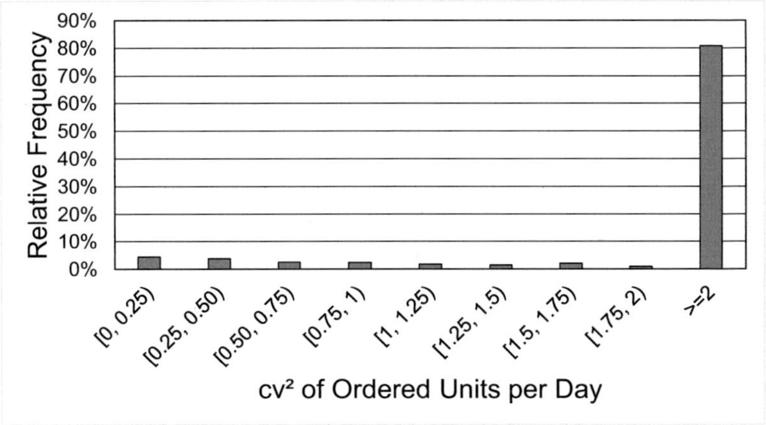

Figure 7.4: Coefficient of variation of order quantities

Figure 7.5 shows the average consumption per day aggregated on basis of the suppliers. We note that most of the suppliers supply an average weight of between 0 and 500 KG/day. The smallest payload typically used in freight forwarding is 3 tons. Therefore this small weight per supplier indicates that there is a need for either temporal or spatial consolidation in order to increase the utilization of the transport means to an acceptable level.

[1] The interested reader may refer to Katsnelson and Kotz (1957), who proved that, for a bounded dataset with N elements all sample values being positive, the squared coefficient of variation cannot exceed N.

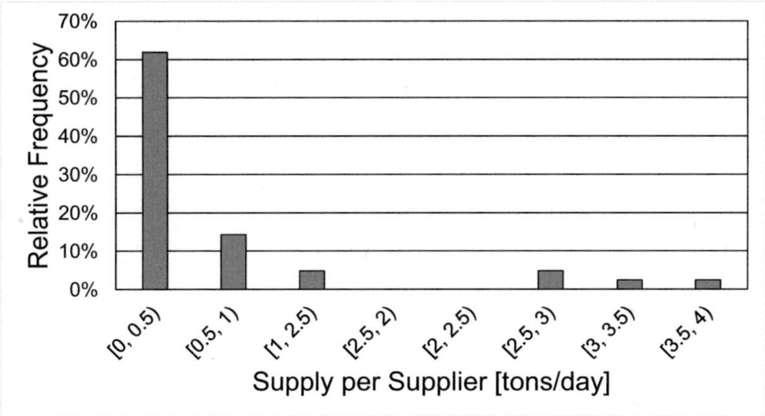

Figure 7.5: Relative frequency distribution of consumption per supplier [tons/day]

Figure 7.6 shows the frequency distribution of the weight per shipment unit. It was calculated as follows:

$$W_{ShipmentUnit} = N_{PartsPerBin} \cdot W_{Part} + W_{Bin} \tag{7.1}$$

We note that most of the shipment units have a size lower than 50kg. These are favorable conditions for leveling. The smaller one unit, the more exact we can adjust the daily weight to the vehicle capacity. The reason is that we can only order discrete unit sizes. The higher the unit sizes, the more excess capacity due to rounding errors will occur. [2]

[2] The theoretical optimum is an infinitely small unit weight, i.e. liquids.

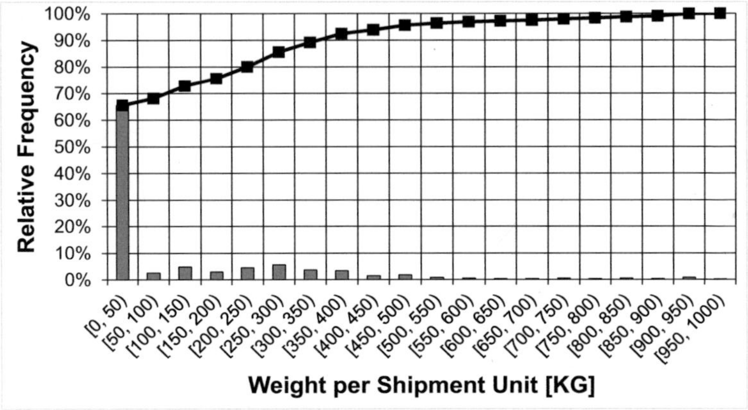

Figure 7.6: Relative frequency distribution of weight per shipment unit

In this chapter we presented the structure of the sample data we were provided and performed some descriptive analysis to describe the data set. This descriptive analysis, indicates that there is potential for improvement by leveled replenishment. First, the shipment unit weights are mostly small which is why they can consolidated to larger shipments without causing excess capacity. Second, there are no large minimum order quantities imposed by the supplier which prevent us from placing small high-frequent orders. In the subsequent section, we will describe how to process the data of our case study in order to design the system.

7.2 Designing the System

In the preceding section we described the sample data as basis of our case study. In this section we describe the steps we follow in the design process.

As presented in section 4.4, we need to answer the following questions in order to the design the system:

- According to which **transport concept** do we need to integrate the supplier in our network?
- Which **suppliers** can be combined to a tour?
- What **truck capacity** is needed for the tour and in what frequency?
- How many **capacity slots** on the truck shall be reserved for each part on each day?
- How much **buffer inventory** is needed for each part to achieve a certain service level?

The subsequent sections provide a detailed description of each of these steps and its results. The first decision we need to take is the transport concept assignment. That is, each supplier in the network must be assigned a transport concept. This will be described in the subsequent section.

7.2.1 Transport Concept Assignment

As stated in section 7.1, the total dataset covers a time span of 18 months, 57 suppliers and 795 parts. For planning our system, we only consider the time span of the first six weeks. All parts, not procured in this time span, are not taken into consideration for our transport concept assignment.

Moreover, we only consider suppliers within a maximum driving distance of four hours from the ZF Friedrichshafen plant in Friedrichshafen. The reason is that legal limitations require a driving time of 8 hours. For a round trip, the maximum driving time therefore is four hours for one way.

For each supplier we calculated the number of suppliers that are reachable from one supplier within a driving time of 90 minutes or less. The calculation was performed for each supplier that remained after the prior steps. All suppliers that had at least one supplier reachable within a driving time of 90 minutes were assigned the milk run concept and are part of our later calculations.

7.2.2 Creation of Tours

After the transport concept have been assigned, the suppliers need to be combined to form tours. We only consider single-staged transports, i.e. direct milk runs without consolidation. As stated in the preceding section, legal limitations require us to form tours with a duration lower than 8 hours. Therefore, the total tour duration must be smaller than 8h. The tour duration consists of the driving time between the tour stops and a loading time for each stop. We assumed the loading time to be a constant of 30 minutes.

To form the tours, we simply perform the sweep algorithm on all the supplier that were assigned the milk run concept in section 7.2.1, starting at quarter to nine and moving clockwise. For each supplier to be integrated in the tour, we perform the driving time check to decide whether the tour is feasible.

Our Strategy is to integrate as many suppliers as possible on each milk run tour. That is, we neglect truck capacity. If the load per supplier was too large to fit on the truck, we would simply increase the frequency of tours.

Figure 7.7 shows the results of our planning efforts. We created five tours, each one containing three to five suppliers. We were not able to form tours consisting of a higher number of suppliers. The reason is the supplier's spatial distribution, i.e. the high distances between the suppliers which result in high driving times.

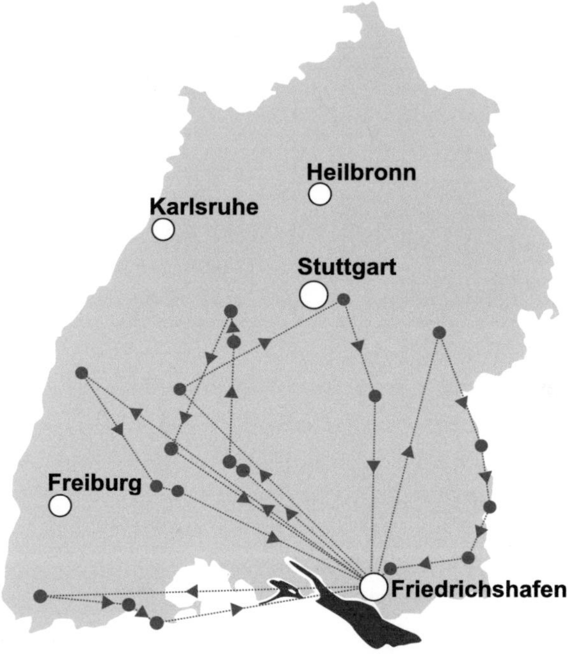

Figure 7.7: Tours of our case study

Table 7.2 summarizes some key figures of the tours that were created. We note that all of the tours have duration close to 8 hours. That is, the available time will be utilized to a high degree. The total weight which will be ordered differs among the suppliers. Whilst in Tour 1 only 5.7 tons will be ordered over the whole leveling horizon, almost 184 tons will be ordered in tour 5. Assuming a minimum truck capacity of 3 tons, we already note that we will need less than twenty transports per month in case of tour 1. We also note that the average weight per shipment unit varies by a factor of more than twenty: in case of tour 3, the average weigh per shipment unit is 9 KG. In case of tour 2, the average weight per shipment unit is 189 KG.

Table 7.2: Basic data of the tours of the numerical case study

	Tour 1	Tour 2	Tour 3	Tour 4	Tour 5
Total Distance [KM]	413	411	422	417	350
Total Duration [h]	7.41	7.38	8.03	7.95	7.46
Number Of Suppliers	3	3	4	4	5
Total Weight in Leveling Period [tons]	5.7	60	24	37.6	183.9
Number of different Part Numbers	21	47	76	113	31
Average Weight Per Shipment Unit [KG]	30.86	189.36	9.01	71.18	110.42

After the tours have been defined we are able to create a leveling pattern. For each tour, this pattern contains information regarding which part is to be procured on which day and in which quantity.

7.2.3 Calculation of a Leveling Pattern

On the basis of the tours we created in the preceding section, we create leveling patterns for the trucks that perform the tours in this section. The leveling pattern contains information regarding which part can be picked up on which day and in which quantity. To create each pattern, we employ the optimization model we presented in section 4.4.3.

As input parameters to create the constraints of the model, we need the total number of units which have to be transported over the leveling horizon, the capacity of the vehicle that performs the transportation and the number of tours that need to be transformed.

The total number of units can be calculated from the average demand per day using formula (4.26). In the base case (see section 7.3) we assume a minimum catch-up capacity of 15%. In section 7.4 we perform a variation of this parameter to find the cost-minimal buffer allocation. If we know the total number of units which need to be transported for each part number, we also know the total weight which needs to be transported over the whole leveling horizon. Therefore we can take a decision regarding which vehicle is to perform the transports and in which frequency. To determine the vehicle capacity and the number of tours, we follow two different Strategies (cf. section 6.2.1).

Strategy A: High frequency and low payload. We assume a target delivery frequency of 20 deliveries per month and select the smallest vehicle size sufficient to perform the transports. If the vehicle size is already at the minimum of three tons, we decrease the number of tours if possible.

Strategy B: Low frequency and high payload. We assume a vehicle with a payload of 20 tons and select the smallest possible frequency. If at the same frequency, a smaller vehicle is possible, select a smaller vehicle.

Following Wilken (2017), truck sizes (i.e. payload) are assumed to be integers between 3 tons and 20 tons. Moreover we assume that the truck size is the same for each day, which is common for the case of milk runs (cf. Meyer 2015). Thus, to determine the truck sizes, we can divide the total weight in the leveling horizon by the number of transports and round up to the next integer.

Figure 7.8 shows the results of the optimization procedure for all tours of our case study and both of the above strategies. We note that we only perform 20 transports in case of tours 2 and 5 and Strategy A. In all further cases, the capacity requirement is too low to perform 20 tours with an acceptable vehicle utilization. Moreover we note that the variability of the total weight which needs to be transported is rather low.

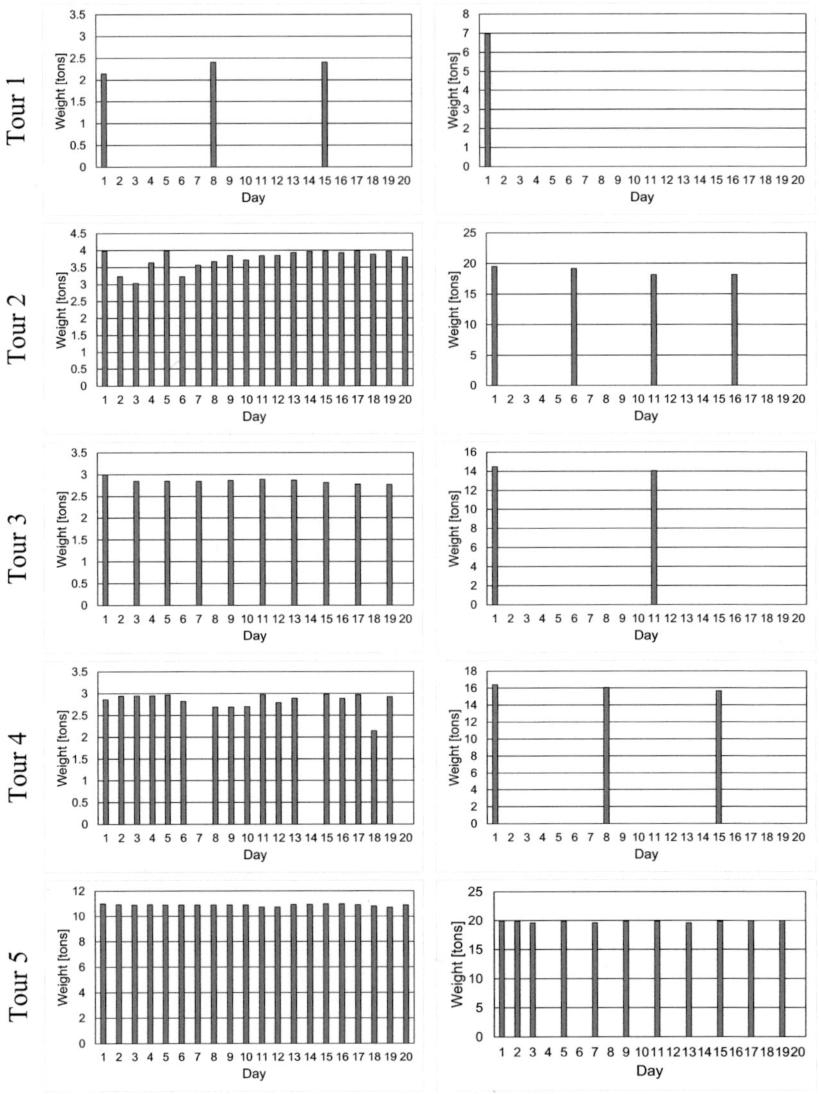

Strategy A: High Frequency

Strategy B: Big Vehicle

Figure 7.8: Total weight per day according to the leveling pattern generated by the optimization model

Tour 1 is the tour with the lowest monthly throughput requirement in terms of weight. That is why both the number of tours and the vehicle size is lowest for this tour. Under consideration of the required catch-up capacity of 15%, a transport capacity of a little less than 7 tons per month is required. In case of Strategy B, the optimization model therefore selected a vehicle with a payload of 7 tons. Since the smallest possible vehicle size is 3 tons, a total capacity of 9 tons needs to be reserved, which results in an excess capacity. That is, capacity which is not being reserved for any of the products and thus will not be utilized.

Tour 2 has the second highest capacity requirement. In case of Strategy A, which aims at achieving the highest possible frequency, a vehicle size of 4 tons with 20 transports per month is selected. In case of Strategy B, 4 transports with a vehicle size of 20 tons are performed.

Tour 3 has the second lowest capacity requirement. In case of Strategy A, 10 tours with a capacity of 3 tons are performed. In case of Strategy B, only two tours with a vehicle size of 15 tons are performed. The capacity requirement of tour 4 is similar to tour 3. For Strategy A, the vehicle payload is 3 tons and 17 tours are performed. In case of Strategy B, 3 tours with a vehicle payload of 17 tons are performed.

Tour 5 is the tour with the highest capacity requirement. For Strategies A and B, 20 tours with 11 tons or 11 tours with 20 tons are performed, respectively.

Figure 7.9 displays the capacity utilization of the truck for the tours of our case study. We differentiate the capacity by three different categories. The average utilized capacity denotes the weight of the exact expected demand without considering catch-up capacity. We want to ensure a catch-up capacity of at least 15% for each part. Since we can only reserve discrete truck sizes, we need to round the next truck sizes. The capacity which is neither required capacity nor catch-up capacity is called excess capacity.

We note that the percentage of excess capacity is the smallest for tour 5. The reason is that we can only choose among discrete truck sizes and need to round the next integer. The percentage of the rounding value decreases, the higher the required vehicle capacity gets. Since tour 5 is the tour with highest throughput, the rounding effect is smaller compared to the other tours.

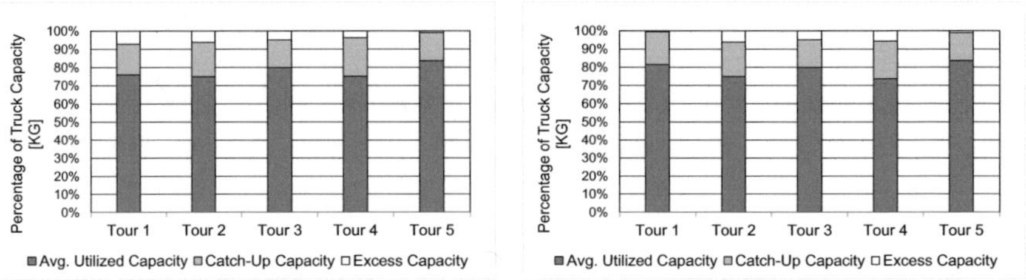

Strategy A: High Frequency

Strategy B: Big Vehicle

Figure 7.9: Vehicle utilization for Strategy A and Strategy B

After the calculation of the leveling patterns for each tour we are able to calculate the inter-arrival times between the orders and hence calculate the inventory buffer necessary to achieve a certain service level. The subsequent section yields a description of the steps we followed in this procedure.

7.2.4 Calculation of Buffer Inventory

Given the leveling pattern as an input factor, the inventory level which is needed in order to buffer against the variations of demand can be calculated. To perform the calculations, we employ the GG1 inventory model in discrete time which was presented in section 4.3.1 and follow the procedure from section 4.4.4.

In our case study, daily demand is a varying random variable. For the replenishment process, the extent of variation is limited. Further on, demand can occur in every period while replenishment only occurs in periods and quantities which are specified in the delivery schedule. The inventory of each period derives from the superposition of the consumption and replenishment process (cf. section 4.3.1).

The five tours of our case study encompass a total of 288 parts. For all parts we employ the same method to calculate the buffer inventory. In the following, we explain the procedure we follow by the example of one single part of Tour 5. Table 7.3 shows the design parameters of our example. The average demand per day is 1.8261 units. Multiplied by twenty days, this yields 36.5 units for the whole leveling horizon. If we round this value up to the next integer, we cannot ensure our minimum catch-up capacity of 15%. To calculate the number of slots we need to reserve, we divide 36.5 by 0.85 and round up to the next integer, which yields a total of 43 units.

Table 7.3: Design parameters for the buffer inventory calculation

Design Parameter	Value
Demand per Day [units/day]	1.8261
Demand in Leveling Horizon [units]	36.5220
Catch-Up Capacity [%]	15
Reserved Capacity Slots [units]	43

Given the reserved capacity slots for our example part number and the remaining part numbers of the tour, we can calculate a heijunka pattern for the milk run tour.

Figure 7.10 shows the heijunka pattern of our example part. On most of the days we are able to order three units. There are some days on which we are allowed to order zero or one unit. The longest time interval between two orders is one period. There are only five periods in which we reserved a capacity of zero units.

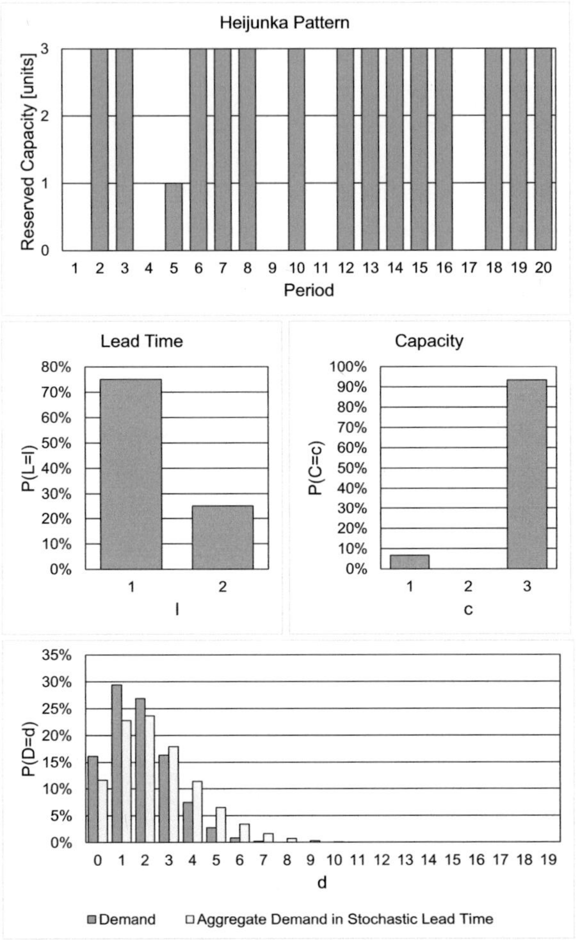

Figure 7.10: Deriving the lead time and capacity distribution from the heijunka pattern

Given the heijunka pattern, we calculate a lead time distribution and a capacity distribution. Note that in the capacity distribution we only consider periods in which the reserved capacity is greater than zero units. The zero-periods are already incorporated in our lead time distribution.

Given the lead time distribution and the probability distribution of the demand per day, we can calculate the aggregate demand in the stochastic lead time.

Figure 7.11 displays the inputs and outputs of the discrete time G|G|1 inventory model.

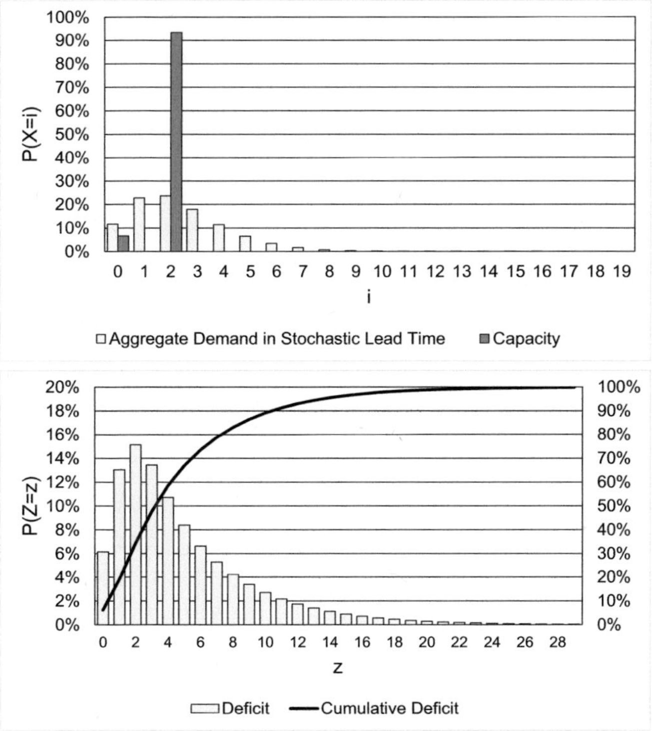

Figure 7.11: Input and output parameters of the G|G|1 inventory model

The top displays the input figures, i.e. the probability distribution of the reserved capacity and the probability distribution of the aggregate demand in the stochastic lead time. From these inputs we can calculate the probability distribution of the deficit and determine the number of kanbans we need for a specified statistical safety. In our case we need a statistical safety of 99%, which corresponds to a total of 21 kanbans.

The dimensioning of the inventory buffer, as performed in this section, is the final planning step of the design of our heijunka-leveled replenishment system. We are now able to calculate the costs which incur due to holding inventory and transport the goods from the suppliers to the receiving plants.

7.3 Calculating the System Operating Costs

Based on the system design described in the preceding section, we will now calculate the system's operating costs. The costs that are considered are transport costs and inventory costs.

To calculate the transport costs, we take an activity-based-costing approach. According to Wilken (2017), transport costs consist of a time- and distance-dependent component. The time component accounts for the wages that are paid to the driver. In a high-wage country such as Germany, this is the major part of the costs. The distance-dependent component to a large extent accounts for the vehicle's gas consumption. Both components depend on the vehicle size, i.e. the higher the payload, the higher the hourly cost and the higher the distance based cost. Another factor that is taken into account is the depreciation due to wear and tear. It also increases with the vehicle size (see also section 6.2.1).

For each tour, the transport cost per month can be calculated as follows:

$$K_{TR} = (t_{Tour} \cdot k_{duration} + d_{Tour} \cdot k_{distance}) \cdot N_T \cdot r_{Markup} \tag{7.2}$$

In the equation, t_{Tour} denotes the duration of the tour, i.e. 8h for all of our tours, and $k_{duration}$ denotes the hourly cost. The distance-dependent cost consists of the total tour distance d_{Tour} and cost per distance unit $k_{distance}$. To calculate the total cost per month, we need to multiply the sum of both components by the number of tours per month N_T which is an input or output factor, depending on the Strategy, of our optimization model. Moreover, we multiply the costs with a factor to add a markup of 15%. This markup rate accounts for administration, risk and a profit margin of the freight forwarder (cf. Meyer 2015).

Table 7.4: Truck cost rates (excerpt from Wilken 2017)

Payload [tons]	Cost per Day [EUR/8h]	Cost per KM [EUR]
5	504.33	0.57
10	536.23	0.68

To calculate the inventory cost of all parts of each tour, we employ our discrete time G|G|1 inventory model. The first step is to calculate the average inventory of each part. For this computation, we calculate the expected value of the physical inventory distribution given by equation (4.15).

Multiplying the expected value of the physical inventory with inventory holding costs yields the inventory costs per part. To calculate the inventory costs of all parts that belong to the tour, we can simply add the inventory costs of the respective parts.

$$K_I = \sum_{i \in \mathcal{P}_{Tour}} E(I_i) \cdot k_{h,i} \tag{7.3}$$

In practice, the inventory cost per part is usually calculated as a percentage of the inventory price per unit. Since unit prices of parts vary, inventory holding costs per unit also vary. In our sample data set, we do not have any information regarding the part's prices. Therefore, we assume a unit price of 0.025 money units per day for each quantity unit. To ensure our results are not falsified regarding wrong assumptions of the inventory holding cost, a sensitivity analysis is performed in the subsequent section.

Given the transport cost and inventory cost, we can calculate the total cost as the sum of both components. Therefore, we need to perform the following calculation for each tour (cf section 6.2):

$$K_{TO} = K_{TR} + K_I \tag{7.4}$$

The results of the analysis are displayed in Figure 7.12 for both Strategies. We note that the largest proportion of the costs are the transport costs. Therefore, the tours with the lowest frequency per month incur the lowest costs. We also note that tour 4 incurs the highest cost of all tours, although its transport costs are lower than the transport costs of tour 5. The reason is that in tour 4, the milk run transports a higher number of number than tour 5. The average unit weight of these parts is lower than the average unit weight of the parts of tour 5. Moreover, the Figure shows that the overall level of costs is lower for Strategy B. The reason is that marginal cost of performing one more tour per month are higher than the marginal cost of choosing a vehicle with a higher payload.

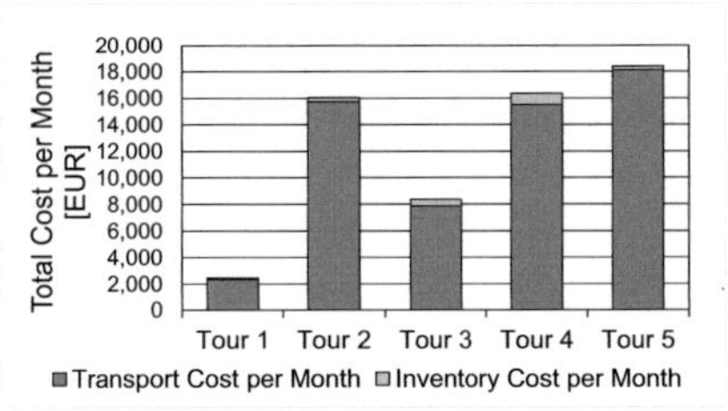

Strategy A: High Frequency

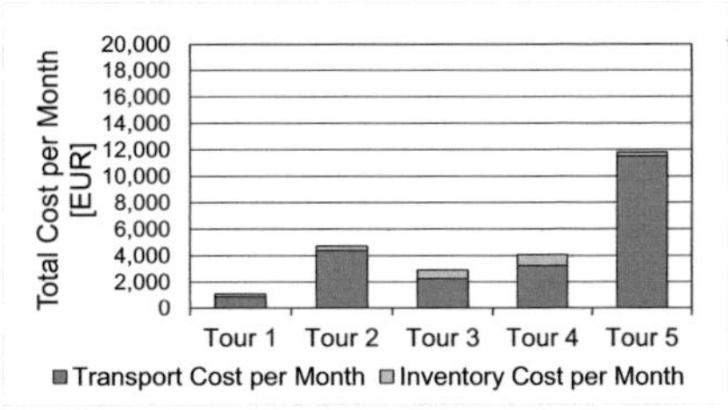

Strategy B: Big Vehicle

Figure 7.12: Operating costs per month of the five tours of our case study

In this section, the five tours of our case study were evaluated from a total cost perspective. In the base-case scenario, we assumed a desired catch-up capacity of 15%. In the following section, we will vary this catch-up capacity to find a point of operation for our system which minimizes the total costs.

7.4 Finding a Cost-Minimal Buffer Allocation

In the preceding section we calculated the operating cost of the system we designed for a fixed given catch-up capacity of 15%. In this section this assumption is relaxed, i.e. we vary the minimum required catch-up capacity and compute the total cost for different buffer allocations. To determine the minimum, we use both a discrete approach, i.e. with discrete truck sizes and a discrete inventory model (cf. sections 4.3.1 and 7.3), and its continuization with continuous truck sizes and a continuous inventory model (cf. section 6.2).

In the first subsection we show how we can determine the optimum buffer allocation by means of our discrete model. In the second subsection, we use the continuous model presented in section 6 to determine the optimum buffer allocation and, in order to evaluate the approximation accuracy, compare its results to the discrete model.

7.4.1 Determining the Optimum Catch-up Capacity with the Discrete Model

In this subsection we calculate the cost of transportation and the cost of inventory by using the transport cost rates from Wilken (2017) with an activity based cost model and the algorithm of Grassman and Jain (1989)(cf. section 7.3). We iterate over different discrete catch-up capacities and create a leveling pattern for each catch-up capacity by using the optimization model presented in section 4.4.3. From the leveling pattern, we extract the required transport capacity to calculate the cost of transportation. Moreover, we extract the quantities and their inter-arrival times for each part to calculate the number of kanbans and the average inventory by using the algorithm of Grassman and Jain (1989) (see sections 4.3.1 and 7.2.4). Given the cost of transportation and the cost of inventory for all catch-up capacities, we can determine the minimum of the cost of operation.

To determine the inventory costs, we assume a rate of 0.025 EUR/day for each part. In practice, the inventory costs vary per part and are proportional to the part value (cf. section 6.2.2). Part weight and part price have an influence on the cost-minimal buffer allocation, e.g. if we an extreme of very low weight but high price parts and vice versa. In order to evaluate the robustness of the results we will obtain in this section, we perform a sensitivity analysis by varying the inventory cost unit rate and investigating its influence on the location of the minimum of total costs. That is, we vary the inventory unit cost rate ceteris paribus and analyze its effect on the location of the cost-minimal buffer allocations. This enables us to hedge against a distortion of the results due inaccurate assumptions regarding the unit cost rate.

The upper half of Figure 7.13 shows the total cost per month as a function the minimum desired catch-up capacity for tour 1 of our case study. For both strategies, the total costs are split in costs which arise due to the operation of the milk run which procures the parts and the cost of average inventory. We observe that at the sides of the horizontal axis, both transport costs and inventory costs increase exponentially (see the preceding section): If we increase or reduce the required catch-up capacity, either transports costs or inventory costs increase exponentially.

We note that the total costs in Strategy B are lower than in case of Strategy A. This is explained by the fact that the larger amount of total costs are caused by the transportation. The transportation cost depend on the size of the vehicle and the number of tours per month. A major part of transportation costs are the wages for the driver. Therefore, reducing the number of transports has a bigger effect on transport costs than reducing the vehicle size.

In case of Strategy A, the optimum catch-up capacity is 30%. In the optimum, we perform 3 tours with a truck payload of 3 tons. In case of Strategy B, the optimum catch-up capacity is 20% at 1 tour with a truck payload of 7 tons. The total cost is 36% lower in case of Strategy B.

The results of our sensitivity analysis for tour 1 in case of both Strategies are also depicted in Figure 7.13. We varied the inventory unit cost rate from a minimum 0.001 EUR/day to a maximum of 10 EUR/Day, i.e. we increase the minimum by a factor of 10,000. We observe that the minimum catch-up capacity increased with the inventory unit cost rate. The increase is not linear but logarithmic. In case of Strategy A, for example, an increase from a cost rate of 0.001 EUR/Day to 1 EUR/Day, i.e. factor 1,000, resulted in an increase of the optimum catch-up capacity by a factor of 2.67.

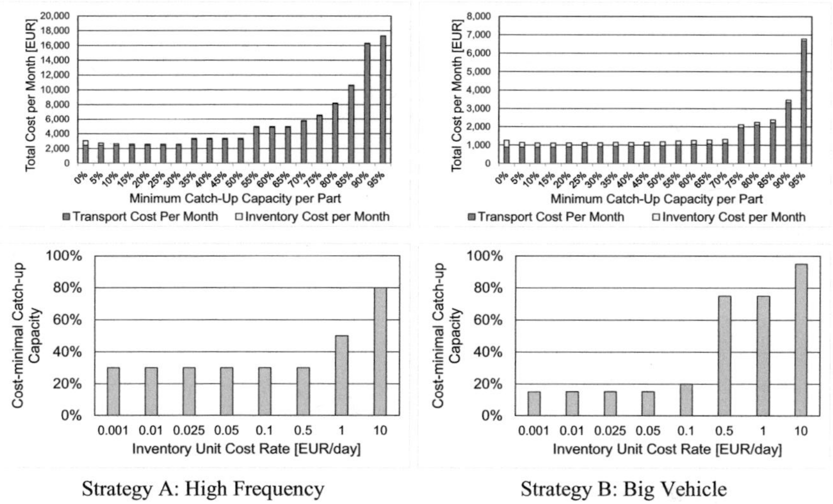

Figure 7.13: Parameter variation of catch-up capacity and sensitivity analysis for tour 1

The upper half of Figure 7.14 shows the results for the second tour of our case study. We note that the sensitivity of transport costs with respect to the catch-up capacity is smaller than at tour 1. The reason is that the low variation of the number of transports. In case of Strategy A, the optimum catch-up capacity is 20%. In the optimum, we perform 20 tours with a truck payload of 4 tons. In case of Strategy B, the optimum catch-up capacity is 15% at 4 tours with a truck payload of 19 tons. The total cost is 72% lower in case of Strategy B. Both optima are located at a step from one vehicle size to the next bigger vehicle.

The lower half of Figure 7.14 displays the results of the sensitivity analysis for tour 2. For Strategy A, an increase of the unit cost rate by the factor of 10,000 leads to an increase of the minimum required catch-up capacity by the factor of 4. For Strategy B, the respective increase is more than by the factor of five. We note that for both Strategies and both tours so far, the location of the optimum is quite robust for unit cost rates of up to 0.1 EUR/Day. The reason is that to a large extent, the optimum is still determined by the transport cost and not inventory cost.

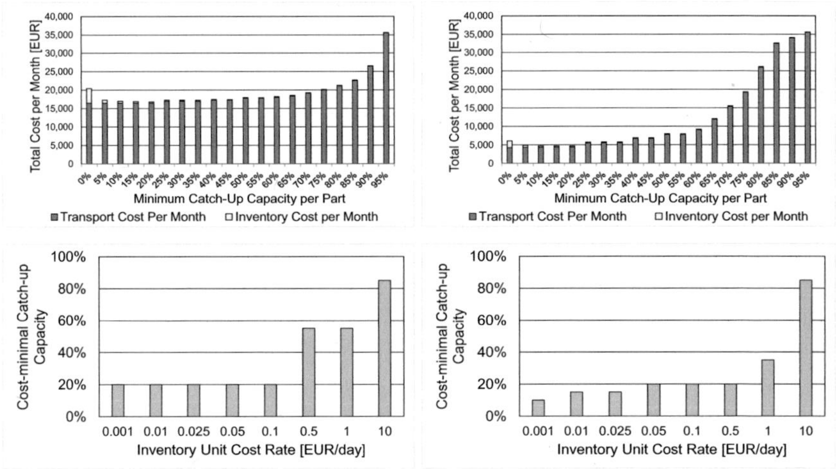

Strategy A: High Frequency | Strategy B: Big Vehicle

Figure 7.14: Parameter variation of catch-up capacity and sensitivity analysis for tour 2

The first row of Figure 7.15 shows the results for the third tour of our case study. The variation of transport costs is higher than at tours two, four and five but lower than at tour one. Again, this is due to increasing number of tours per month with increasing catch-up capacity. In case of Strategy A, the optimum catch-up capacity is 5%. In the optimum, we perform 9 tours with a truck payload of 3 tons. In case of Strategy B, the optimum catch-up capacity is 20% at 2 tours with a truck payload of 20 tons. Both optima are located at the tipping point of the number of transports per month. Since the catch-up capacity in case of Strategy B is higher than in case of Strategy A, the total amount of required capacity ($N_T \cdot C_{Vehicle}$) is higher. However, since the number of tours N_T is smaller, the total costs are lower as well. In the optimum, the total cost are 67% lower in case of Strategy B.

The second row of Figure 7.15 displays the results of the sensitivity analysis for tour 3. For Strategy A, an increase of the unit cost rate by the factor of 10,000 leads to an increase of the minimum required catch-up capacity by the factor of 4. For Strategy B, the respective increase is by the factor of 8.

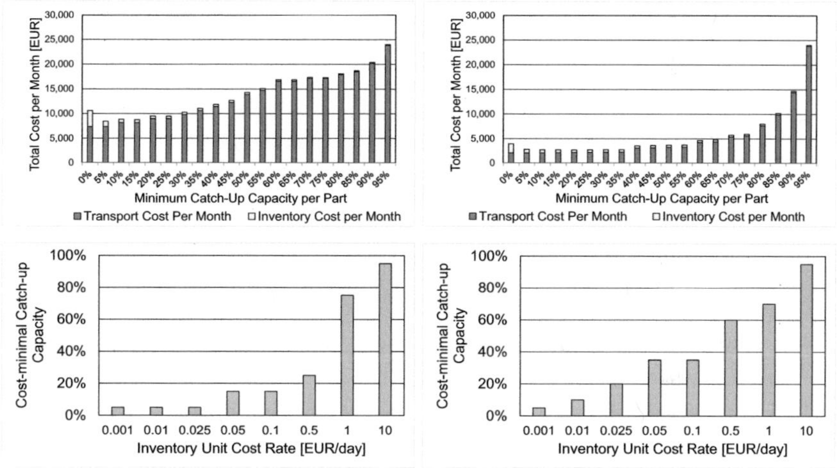

Strategy A: High Frequency

Strategy B: Big Vehicle

Figure 7.15: Parameter variation of catch-up capacity sensitivity analysis for tour 3

Figure 7.16 shows the results for the fourth tour of our case study. In case of Strategy A, the optimum catch-up capacity is 15%. In the optimum, we perform 17 tours with a truck payload of 3 tons. The optimum is located the tipping-point from one vehicle size to a lower one. In case of Strategy B, the optimum catch-up capacity is also 20% at 3 tours with a truck payload of 17 tons. Similar to tour 3, the catch-up capacity is higher in case of Strategy B, which leads to a higher overall required capacity. Again, the total costs are lower for Strategy B because less tours are performed per month. The total cost is 72% lower in case of Strategy B.

The sensitivity analysis of tour 4, depicted in Figure 7.16, shows that an increase of a factor 10 yields a catch-up capacity twice as high as the original one and an increase by a factor of 100 yields a catch-up capacity the nineteen-fold of the original one.

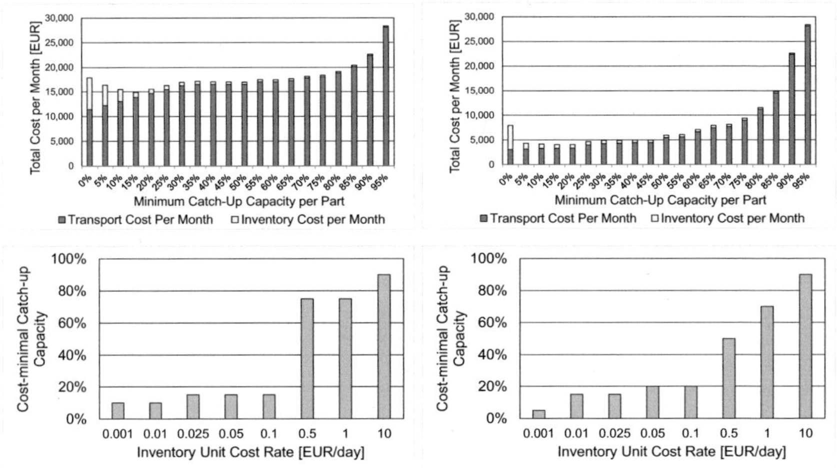

Strategy A: High Frequency   Strategy B: Big Vehicle

Figure 7.16: Parameter variation of catch-up capacity sensitivity analysis for tour 4

The top of Figure 7.17 displays the results for tour five. As stated before, the demand for transport capacity, throughput [KG/Day], is highest in tour five. Therefore, the relative share of inventory cost is lowest for this tour. In case of Strategy A, the optimum catch-up capacity is 15%. In the optimum, we perform 20 tours with a truck payload of 11 tons. The optimum is located the tipping-point from one vehicle size to a lower one. In case of Strategy B, the optimum catch-up capacity is 5% at 11 tours with a truck payload of 20 tons. It is located at the tipping point of the number of transports per month. The total costs are lower for Strategy B because less tours per month are needed. The total cost is 39% lower in case of Strategy B.

The sensitivity analysis of tour 5 is depicted in the lower half of Figure 7.17. For Strategy A, an increase of the unit cost rate by the factor of 10,000 leads to an increase of the minimum required catch-up capacity by the factor of 6. For Strategy B, the minimum required catch-up capacity is 0% for an inventory cost rate of 0.001.

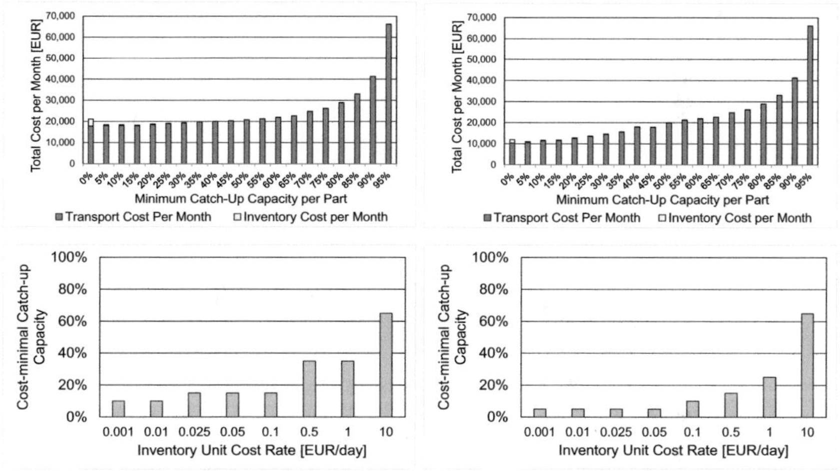

Strategy A: High Frequency

Strategy B: Big Vehicle

Figure 7.17: Parameter variation of catch-up capacity sensitivity analysis for tour 5

Figure 7.18 depicts the total cost of all five tours for the two different Strategies. We note that in our case study, Strategy B which chooses a (large) truck size and adjusting the delivery frequency (LF) was superior to Strategy A, i.e. defining a (high) delivery frequency and choosing a (small) truck size (HF). The reason is the total costs are to a large extent determined by the transport costs. The transport costs are, as stated in section 6.2.1 to a large extent determined by fixed costs. Therefore, saving a tour per month has a greater effect on the costs than choosing a smaller vehicle and keep the number of tours constant. Since Strategy B aims at minimizing the number of tours, the total costs are lower than in case of Strategy A. On the other hand the resulting lower delivery frequency leads to higher costs of buffer inventory, but the effect is being compensated by the savings in transport costs.

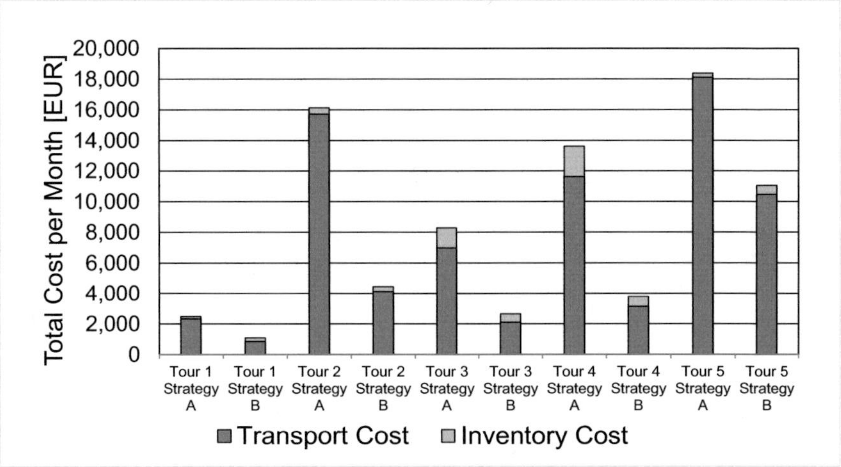

Figure 7.18: Total costs per month for Strategies A and B

Table 7.5 summarizes the results we obtained in this section. We note that in most of the observations, the optimum catch-up capacity was at around 15%. In case of tour 1 and Strategy A, the optimum CUC is at 30%. The reason is that the total weight we need to transport is about 7 tons/month. We provide a capacity of 9 tons. Since the minimum truck size is 3 tons, we can neither reduce the number of transport nor the truck sizes, which leads to this rather high catch-up capacity. The same happens at tour 3 and tour 4 in case of Strategy B.

Table 7.5: Cost-minimal buffer allocations of our case study – Summary of results

		Tour				
		1	2	3	4	5
Strategy A Highest Frequency	Tours per Month	3	20	9	17	20
	Truck Capacity [tons]	3	4	3	3	11
	Optimum Catch-Up Capacity	30%	20%	5%	15%	15%
Strategy B Biggest Vehicle	Tours per Month	1	4	2	3	10
	Truck Capacity [tons]	7	19	16	19	19
	Optimum Catch-Up Capacity	15%	20%	20%	15%	5%

Since the inventory unit cost rate is based on an assumption, a sensitivity analysis to test the robustness of our results was performed for each tour. We found that changing the inventory unit cost rate affected the location of the optimum in a logarithmic manner. This indicates a high robustness of the results.

A general observation which is true for all the tours of our case study is that the cost curve is rather flat around the optimum. That means, the elasticity of the cost curve with respect to catch-up capacity is low and a change in the catch-up capacity only has a relatively small impact on the costs. A managerial insight that can be drawn from that, is that in practice, other factors which are exogenous to our model, might play a role in determining the right catch-up capacity. Consider the following example: the optimum catch-up capacity requires a buffer stock which can only be held if a certain investment in a bigger rack is made. In this case, it might be beneficial to operate the system at a higher (non-optimal) catch-up and hence a smaller buffer stock, because the investment costs can be saved.

7.4.2 Evaluation of Approximation Accuracy

In the preceding section, we used the discrete model of Grassman and Jain (1989) to calculate the system operating cost and determine the minimum of the total cost function. In this section, we will also calculate the optimum catch-up capacity by means of the continuous model, based on Gudehus (1976), we presented in section 6.3 in order to evaluate its approximation accuracy. Therefore, we will compare both the results which can be obtained by the enumeration of the problem (c.f. section 6.3.1) and by the simplified problem, which was linearized by means of a Taylor series (c.f. section 6.3.2), with the results of the discrete model.

Table 7.6 shows both the input parameters we used and the results we obtained from the calculation of the different models. The factor costs of inventory κ_I were obtained by multiplying the holding costs per unit k_h with the number of parts N_P (see Table 7.2 and section 6). The variability parameter f of the continuous time G|G|1 model was calculated as the arithmetic mean of the variability parameters over all parts (c.f. 0). For the sake of simplicity, we assumed $v_a^2 = 0$ for each parts instead of calculating the actual coefficient of variation of the leveling pattern. As coefficient of variation of the service time v_b^2, we simply used the coefficient of variation of demand.

In the calculation of κ_T we differentiated between the two planning Strategies A and B (see section 6.2.1). For Tours 1, 3 and 4 of Strategy A, we used (6.27) to determine κ_T. For Tours 2 and 4, we used (6.28). In case of Strategy B, we used (6.29) for all tours.

In case of Strategy A, we note that the accuracy of the continuous model was highest in case of tour 5. The reason is that in tour 5, the total throughput in terms of weight per month is relatively high and the number of parts and hence the inventory costs is relatively low. Therefore, the ratio of transport costs to inventory costs is higher than for the other tours. Therefore, the relative error which is caused by approximating the average inventory of the discrete model by the average deficit of the continuous model is relatively small, leading to higher overall accuracy.

Moreover, we note that the catch-up capacity which was calculated by the continuous model for tours 3 and 4, is about 6% higher than the catch-up capacity that was determined by the discrete model. The reason is that the continuous model systematically overestimates the average inventory. For both tours 3 and 4, the monthly throughput is relatively low whereas a high number of different parts is transported. Therefore, the relative impact of inventory on the total operating cost is higher compared to the other tours. This distorts the result and leads to a higher catch-up capacity.

The largest error can be observed in case of tour 1. According to the discrete model, the optimum catch-up capacity is 30%, whereas the continuous model yields a catch-up capacity of only 6%. The reason for this deviation is the fact that transport costs are a step function in case of the discrete model and a linear function in case of the continuous model. In the discrete model, the catch-up capacity can be increased without any change in transport costs. At the transition from a catch-up capacity of 30% to a catch-up capacity of 35%, an extra tour is required which results in an increase in cost. Therefore, the optimum is at 30%. In case of the continuous model, we assume a continuous truck size. The optimum is located where the increase in cost of transportation is equal to the increase in cost of inventory. This point is reached at a catch-up capacity of 6%.

The second largest error can be observed in case of tour 3. The discrete model yields an optimum catch-up capacity of 5%, whereas the continuous model yields 11%. The reason is that tour 3 contains many parts which can be described as extreme low runners. The average demand for these parts is so low that we need to reserve capacity for only 1 unit per month. This single unit already covers almost a 100% of catch-up capacity. The costs caused by the remaining parts are not sufficient to offset the transport costs to shift the optimum to a higher catch-up capacity. The continuous model, on the other hand, does not allow for this differentiation. It assumes that all parts react in the same way on a change in catch-up capacity and simply multiplies the effect by the number of parts. Therefore, the impact of inventory on total costs gets overestimated, resulting in a higher optimum catch-up capacity.

Table 7.6: Comparison of optimum catch-up capacity calculated by different models

			Tour 1	Tour 2	Tour 3	Tour 4	Tour 5
Input Parameters	General	κ_I	15.75	35.25	57.00	84.75	23.25
		N_{Parts}	21	47	76	113	31
		f	-1.14	-3.44	-2.84	-3.34	-1.33
	Strategy A	κ_T	1,530.13	1,020.60	6,431.09	10,073.97	2,680.47
		κ_T/κ_I	97.15	26.9	112.83	118.87	115.29
	Strategy B	κ_T	266.27	2,798.21	1,119.11	1,753.01	8,174.37
		κ_T/κ_I	16.66	78.24	19.35	20.39	346.53
Results (CUC_{opt})	Strategy A	CM	6.00%	16.00%	11.00%	11.00%	14.00%
		CM-T	12.08%	14.52%	13.03%	13.18%	14.80%
		DM	30.00%	20.00%	5.00%	15.00%	15.00%
	Strategy B	CM	16.00%	11.00%	22.00%	24.00%	9.00%
		CM-T	14.44%	13.78%	14.61%	14.64%	6.39%
		DM	15.00%	15.00%	20.00%	15.00%	5.00%
CM	Continuous Model (Gudehus)						
CM-T	Continuous Model – Taylor Approximation						
DM	Discrete Model (Grassman and Jain)						

In case of Strategy B, we note for tours 1 and 2 that both the continuous model and its Taylor series approximation yield concurrent results with a catch-up capacity of about 10-15%. In case of tour 3, the continuous model is more accurate than its Taylor series approximation, its result yielding only an absolute difference of 2% to the discrete model. In case of tour 4, the opposite is the case: The result which was obtained with the Taylor series approximation is closer to the result of the discrete model than the enumerated continuous model. Again we note that for tours 3 and 4, the optimum catch-up capacity is overestimated by the continuous model due to the high number of parts. In case of tour 4 this effect is mitigated by the linearization, resulting in a more accurate result of the Taylor series approximation of the continuous model compared to the actual model. In case of tour 5, the results of the continuous model, its approximation, and the discrete model are concurrent. The difference can, again, be explained by the linearization of - actually discrete - vehicle sizes.

In some cases, the simplified continuous models yield a catch-up capacity which deviates from the exact catch-up capacity which was calculated by means of the discrete model. If we would operate the system at these points of operation, an allocation inefficiency or deadweight loss occurs (cf. section 6.1).

Table 7.7 shows the allocation inefficiency for all five tours of our case study and both planning strategies we considered. To obtain the allocation inefficiency, we used the cost function of the discrete model and evaluated its value at the catch-up capacities given in Table 7.6. Since we only know the values of the discrete cost function at discrete points with an interval size of 5%, we rounded respectively to determine the costs.

The allocation inefficiency is quantified as the relative deviation between the total cost function's value at the optimum catch-up capacity obtained by the discrete model and the total cost function's value at the non-optimal catch-up capacity calculated by the continuous model. It can be expressed as:

$$\Delta_{rel}(CUC_{CM}) = \frac{K_{TO}(CUC_{CM}) - K_{TO}(CUC_{DM})}{K_{TO}(CUC_{DM})} \tag{7.5}$$

We note that the allocation inefficiency caused by the inaccuracies of the continuous models is relatively low for the tours of our case study. All relative deviations are below 10%. With the exception of tour 1 in case of Strategy A, all relative deviations are below 5%. This means that the increase in total costs per month incurred by the miscalculation of optimum catch-up capacity is below 5%. The reason is that, as stated in the preceding section, the cost curve is flat around the optimum.

Table 7.7: Allocation inefficiency caused by the approximation error of the continuous models

			Tour 1	Tour 2	Tour 3	Tour 4	Tour 5
Strategy A	Costs [EUR]	$K_{TO}(CUC_{CM})$	2,750	16,899	8,864	15,482	18,248
		$K_{TO}(CUC_{CM-T})$	2,655	16,899	8,777	14,866	18,248
		$K_{TO}(CUC_{DM})$	2,575	16,782	8,464	14,866	18,248
	Deviation [%]	$\Delta_{rel}(CUC_{CM}$	6.80%	0.70%	4.73%	4.14%	0.00%
		$\Delta_{rel}(CUC_{CM-T})$	3.11%	0.70%	3.70%	0.00%	0.00%
Strategy B	Costs [EUR]	$K_{TO}(CUC_{CM})$	1,112	4,715	2,730	4,667	11,696
		$K_{TO}(CUC_{CM-T})$	1,112	4,706	2,745	4,067	11,040
		$K_{TO}(CUC_{DM})$	1,112	4,706	2,730	4,067	11,040
	Deviation [%]	$\Delta_{rel}(CUC_{CM}$	0.00%	0.05%	0.00%	4.04%	3.59%
		$\Delta_{rel}(CUC_{CM-T})$	0.00%	0.00%	0.18%	0.00%	0.00%

In this section we compared the optimum catch-up capacities that were predicted by the continuous model of section 6.2 with the results of the discrete model. The results look promising. In general, the results show a high accuracy and only few deviations. Some of these deviations can be explained by the fact the continuous model assumed vehicle sizes to be continuous although they are in a fact a discontinuous step function. Moreover, we noted that the average inventory gets overestimated by the continuous model. The distortion which is created by this effect, is relatively low in this particular numerical example, because the inventory costs are low. Taking into account the fact that the curve of total system operating costs is rather flat around the optimum (see section 7.4.1), we note that the deadweight loss or allocation inefficiency which is caused by reserving slightly more or less catch-up capacity, is only marginal. Therefore we can recommend the continuous model to calculate an appropriate amount of catch-up capacity. Although the method is not as accurate as the discrete model, it is much less effortful to obtain the results.

7.4.3 Conclusion

In this section, we used two different methods to determine the cost-minimal buffer allocation for the five tours of our case study in case of two different planning strategies. In the first subsection, we assumed discrete vehicle sizes and used a discrete inventory model to calculate the system operating costs. By enumerating over a discrete axis and comparing the values of the total cost function, we determined the minimum. In order to hedge against inaccurate assumptions regarding the inventory unit cost rate we performed a sensitivity analysis. The results show that the location of the optimum is a logarithmic function of the inventory unit cost rate, which indicates a high robustness.

In all tours of the case study, Strategy B was considerably superior to Strategy A in terms of total cost of operation. There are, however, some limitations to our study which is why a universal recommended action regarding the superiority of the second Strategy cannot be derived. One limitation is that we only calculated inventory holding costs, not inventory-related investment costs. The holding costs encompass interest cost which are caused by the tied capital, the costs for the physical handling of the units and average costs for obsolescence. In practice, the amount of holding costs is relatively low. If a reliable high-frequency delivery enables the operator to leave out an entire supply chain buffer tier, e.g. a larger warehouse, procurement in a small truck might be superior to procurement in a large truck.

Another limitation is that in practice, inventory unit cost rates are a function of the part price and the inventory holding interest rate. In our example, this was not considered. That is, in our example, all parts have same price, which is an invalid assumption. In order to mitigate this limitation a sensitivity analysis was performed. The results indicate a low sensitivity regarding the inventory unit cost rate. This is why this limitation can be considered as less severe than neglecting inventory-related investments.

In the second subsection, we used a simplified continuous model as an approximation of the discrete model to determine the optimum catch-up capacity. Although we observed deviations, the results of the continuous model were concurrent with the results of the discrete model. The deviations could be explained by errors due to the continuous consideration of de-facto discrete vehicle sizes and an over-estimation of buffer inventory due to the inventory model. Moreover, we noted that the allocation inefficiency which was encountered due to misestimating the Pareto-efficient catch-up capacity was marginal.

One important observation of this numerical case study was that the cost-curve is rather flat around the optimum. For practitioners this means that when designing a transport logistics system, one should not only simply calculate the Pareto-efficient catch-up capacity and choose it as the point of operation. The selection of the catch-up capacity might also influence other factors which are not considered in our model. Therefore, to find a global optimum, a certain range of points near the Pareto-efficient point should be considered as the point of operation (cf. section 7.4.1).

The results of the quantitative investigation support the hypothesis that there is a buffer-cost-minimal catch-up capacity in between the two extremes of an all-inventory or an all-capacity buffer. The position of this optimum catch-up capacity depends on part weights and part prices. In our case study from the German automotive industry, this value is between 5% and 30%. That means that buffering the variability with inventory is associated with lower costs than buffering the variability only with reserve transport capacity.

The law of variability buffering states that variability must be buffered by some combination of capacity, inventory or time (cf. section 2.4). Therefore, if we not explicitly take a decision regarding the reserved capacity, we actually take a decision for a capacity buffer and against an inventory buffer[3]. This decision is not wrong per se. In some cases, low-priced capacities or small/lightweight parts with little capacity requirements this might even be the best choice – but we cannot be sure about that. If we follow the steps and methods that we

[3] Or, in case of an insufficient capacity buffer: a decision for not fulfilling the customer's requirements.

proposed in this section, we can find the optimum buffer allocation and set it as the system's point of operation.

7.5 Comparison to Area Freight Forwarder

In the preceding section, the total operating costs as the sum of inventory costs and transport costs were calculated and an efficient point of operation was determined. In this section, we compare the transport costs in the optima of the preceding section with the costs that would incur with an MRP-controlled replenishment and an Area Freight Forwarder.

The analysis addresses different questions:

- What are the costs of transportation in case of the **MRP**-created delivery schedule and an **area freight forwarder**?
- What are the costs of transportation in case of the **heijunka-leveled pattern** and an **area freight forwarder**?
- How high are these costs in comparison to the **heijunka leveled pattern and a milk run**?

For our comparison, we first have to create an area freight forwarder tariff table. The step is necessary since AFF tariff tables are not available in public. Therefore, in the first section, we describe the steps we need to follow to create the tariff table. Afterwards, we explain the results.

7.5.1 Creation of AFF Tariff Table

An area freight forwarder tariff table yields a certain monetary charge for a given (distance, shipment weight) tuple. These are used by freight forwarders to charge the transport services they provide to the consignee. The shipment size is measured as shipment weight and matched with a weight class in the tariff table.

Figure 7.19 displays the tariff table provided by Wilken (2017). The table contains prices for shipment sizes larger than three tons. Since this table provides an orientation how much freight forwarders should charge in order to cover their expenses, we still need to add a markup rate which is charged in order to create a profit.

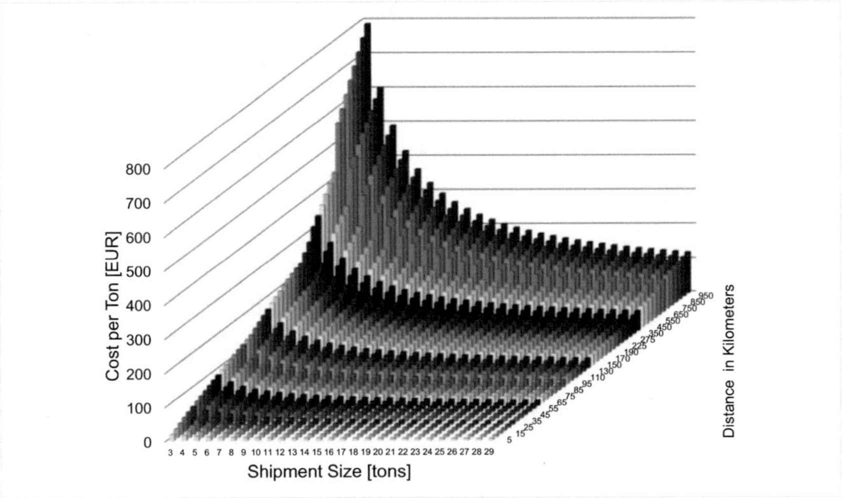

Figure 7.19: Tariff table from Wilken (2017)

To create the AFF tariff table, the tariff table given by Wilken (2017) is extended by adding weight classes for shipment sizes smaller than three tons. The weight classes were provided by Locom Software GmbH.

For our tariff model we calculated the cost for smaller shipments proportional to costs for the three ton truck and added a markup rate. This was repeated for all distance classes of the table. We assumed a markup rate which decreases linearly with increasing shipment size. That is, for shipment sizes between 0 KG and 50 KG, the markup rate is 30%. At a shipment size of three tons and greater, the markup is 15%. This corresponds to the markup rate used by Meyer (2015).

The extra markup is charged by the freight forwarder for the risk of poor vehicle utilization. Since, as pointed out before, the cost performing a transport service is largely fixed and does not depend on shipment size, the profitability depends to a large extent on the overall vehicle utilization. In case of small shipment size, this risk is higher than for bigger shipments. To hedge against this risk, the freight forwarder charges the surcharge.

Figure 7.20 displays the result of our tariff estimation for the example of a distance of 100 KM. We note that the slope is almost linear, since the markup rate only has a minor influence regarding the total price.

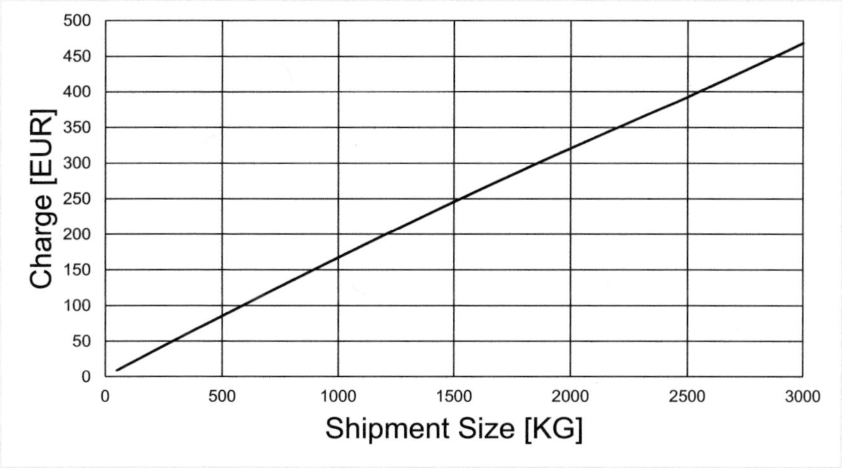

Figure 7.20: Area freight forwarder tariff table for shipment sizes smaller than 3 tons. Distance = 100KM

To calculate the transport costs according to the tariff table, we calculate the total weight which is transported each day for each supplier. This weight is multiplied by the distance to ZF. The total costs per tour are calculated by adding the transport costs of each supplier over our period of investigation of one month.

7.5.2 Results

Figure 7.21 shows the results of our comparison between the different transport concepts, control policies and Strategies.

We observe that heijunka in association with the Strategy which aims at achieving a low frequency with a large vehicle is superior, i.e. lower in respect to transport cost, in all five tours of our case study. In four of the five tours, heijunka with a Strategy which aims at achieving a high frequency incurred the highest transport costs. In three of the five tours, a heijunka leveled replenishment policy in combination with the AFF incurred lower transportation costs than MRP in combination with the AFF.

We emphasize that inventory costs are not included in this calculation. Drawing a conclusion regarding the total costs is therefore not possible. However, since delivery frequencies are the lowest in case of the MRP-controlled replenishments, we suspect that inventory costs are higher in case of MRP.

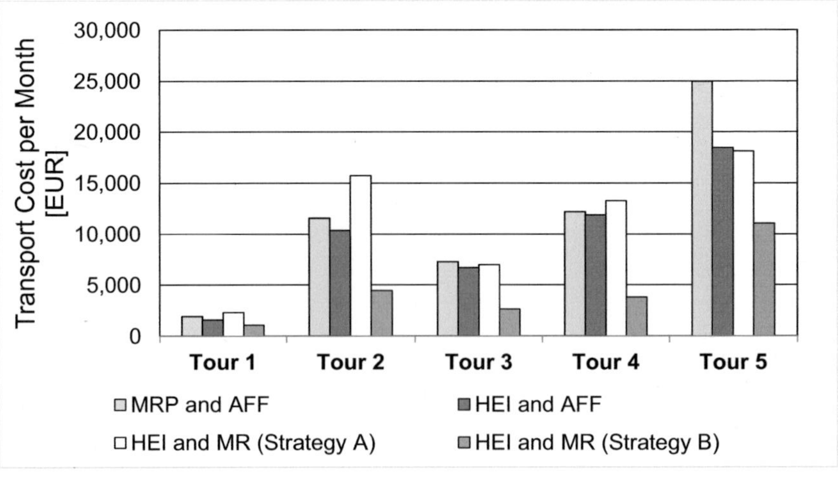

Figure 7.21: Transport costs of different combinations of control policy and transport concept

The computational experiments of this section show that in our case study of ZF Friedrichshafen, heijunka leveling in combination with milk runs is an efficient alternative to the existing system of MRP-controlled replenishment and transports by the area freight forwarder. However, as we pointed out in section 3.1, the milk run has some distinctive advantages in comparison to the area freight forwarder concept.

8 Conclusion and Outlook

The Swedes are no Dutchmen - we saw that very clearly.
(Franz Beckenbauer, German football legend)

Lean Management aims at eliminating the three system inhibitors waste, variability and inflexibility which degrade the system performance. Central elements of Lean Management are a culture of continuous improvement and system design measures which lead to a “stabilization” of processes. Whereas lean techniques are already widespread among industry practitioners in production logistics and enjoy growing popularity in the field of warehousing, the level of maturity is still low regarding systems of transport logistics. This thesis makes a contribution at bringing lean to transport logistics by showing how stabilizing design measures from production logistics can be transferred to transport logistics. The results indicate that current systems of transport logistics could be operated with lower costs and more transparency for the consignee of freight.

8.1 Conclusion

The goal of this thesis was to show how the techniques of lean production can be applied to stabilize transport logistics systems and to evaluate them regarding effectiveness and efficiency.

In Chapter 2, we have reviewed different approaches regarding a definition of stability. Based on a linguistic definition which was given by the dictionary, the requirements to a proper definition of stability were identified. We found that existing definitions lacked certain of these elements. Therefore, we combined certain elements of different approaches to formulate our definition of stability in this thesis as the probability of a process outcome being within a desired target state.

Chapter 3 reviewed the basics of transport logistics systems. The goal was to create an understanding of existing systems and its shortcomings. The first section explained different concepts to handle the physical flow of materials. Afterwards, we presented replenishment policies for controlling the flow information. Based on these elaborations, we investigated how the flow of information and the flow of physical goods can be combined to form transport logistics systems. Further on, the role of stability in systems of transport logistics was pointed out and it was described how stabilization measures offer the potential to improve the system's performance.

In Chapter 4, we described how we can transfer the concept of heijunka leveling from production logistics to transport logistics. We first gave a brief of summary of the functionality of a heijunka-leveled production system. Based on that, we described how the system elements can be moved into a materials supply environment to control the flow of information in a system of transport logistics. Further on, we built mathematical models of the system to understand the system's behavior. Based on the system description and system model, we explained the steps we need to follow to design the system and create a Plan for Every Part.

In Chapter 5 we evaluated the effectiveness of heijunka leveling as stabilization measure. An agent based simulation model was developed and a simulation study investigated 50 different parameter combinations regarding part consumption and weight per part. We found that leveling is effective both in stabilizing the required capacity on the transport level and the orders that are placed at the suppliers on individual part level. We found that the more variable the demand and the more unequal the weight is distributed among parts, the more pronounced is the effect of our leveling policy. The less variable the demand and the more evenly split the weight, the more pronounced is the pooling effect which diminishes the comparative advantage of the leveled replenishment policy in comparison to an unleveled one.

In the sixth chapter we evaluated heijunka leveling in terms of efficiency. First, we explained how the term efficiency is defined according to Pareto and how it can be transferred to the scope of transport logistics systems. Afterwards, the system's operating cost were modeled as a function of the buffer allocation by considering the trade-off between inventory and capacity buffers. Based on this cost function, we derived the marginal cost function and determined the Pareto-efficient buffer allocation as a cost-minimal point of operation. Moreover, we investigated how the location of this optimum changes with varying factor cost ratios. We found that there exists an optimum between an all-inventory and an all-capacity buffer and presented two methods to determine it.

Chapter 7 showed how the methods presented in this work can be applied to a real world example. In a case study, which is based on data provided the German Automotive Supplier ZF Friedrichshafen, we designed a transport logistics system by allocating transport concepts, forming tours, generating heijunka patterns and calculating necessary buffer stocks. Based on the system design, the system operating costs for different buffer allocations were calculated to evaluate the trade-off between inventory and capacity buffers. Furthermore, we compared the operations costs of the heijunka controlled system to the actual system which is controlled by Material Requirements Planning (MRP) in association with an area freight forwarder. We found that in our case study, a heijunka-leveled system with a milk run results in lower operating costs than the MRP-controlled system with an area freight forwarder.

8.2 Outlook

Where can we further improve the design for stability in transport logistics which we proposed in this work? Are there related problems which could benefit from a design for stability?

The optimization model we presented in this work calculates a leveled pattern for a given capacity with the minimal possible order quantities. Limitations regarding capacity can result in patterns which are not inventory-minimal. In some cases, the required transport capacity is less important for the planner than inventory. For these cases, an optimization pattern which yields an inventory-minimal pattern could be developed.

One more potential improvement is related to the inventory model. In the model we proposed, every part is treated as one entity which is separate from all the other parts. That is, we perform leveling on part level. An alternative form is leveling on a family level. In this approach, parts are combined to groups and the capacity is reserved for the group. In each group, there is a variance-reducing pooling effect of the different parts, which could result in a lower required transport capacity.

Moreover, parts with the same supplier could be treated as one family. With this approach, the total number of parts procured from a supplier can be leveled and aligned with the supplier's production capacity. This would reduce the required finished goods-inventory at the supplier. Since different parts are competing for the same capacity slot, we now require a decision rule. This rule changes the replenishment lead time in a way which cannot be considered by current inventory models for leveling on part level. Up to date, there are no analytical models which describe the system behavior of leveling at the family level. In order to be able to determine the necessary buffer inventory, we first need to develop a model for leveling on a family to leverage this potential.

Another application case of an inventory model on family level would be two staged transport networks. With the methods proposed in this work, two consignees who order the same part from one supplier are modeled as two consignees who request two different products. If we had an inventory model to incorporate this constellation, it could be modeled as two consignees who request products from the same family.

Moreover, we think that the trade-off between choosing a buffer of inventory, capacity or time exists in many other different problems of logistics. One further idea could be to adapt to approach of leveling to picking in warehouses. In this environment, leveling can be performed by shifting orders along the time axis and create a leveled workload. The trade-off that exists here is that the order processing costs can be reduced if the customer accepts a certain lead time. The longer this lead time, the less capacity buffer, e.g. extra staff to cover peaks in workload due to express orders, is required.

References

Alicke, K. (2005). Planung und Betrieb von Logistiknetzwerken: Springer.

Arnold, D., & Furmans, K. (2009). Materialfluss in Logistiksystemen (6., erweiterte Aufl.). VDI-Buch. Berlin, Heidelberg: Springer-Verlag Berlin Heidelberg. Retrieved from http://dx.doi.org/10.1007/978-3-642-01405-5

Baudin, M. (2005). Lean logistics: the nuts and bolts of delivering materials and goods: CRC Press.

Branke, J., Häußler, D., & Schmidt, C. (2007). Transport Channel Selection. In Operations Research Proceedings 2006 (pp. 349–354). Springer.

Bronstein, I. N., Mühlig, H., Musiol, G., & Semendjaev, K. A. (2016). Taschenbuch der Mathematik (10. überarbeitete Auflage). Haan, Rhein: Verlag Europa-Lehrmittel Nourney. Retrieved from http://www.europa-lehrmittel.de/t-1/taschenbuch_der_mathematik_bronstein-2657/presseinformation/9783808557891.pdf

Conze, M. H. (2014). Integrierte Abruf-und Transportsteuerung für lagerhaltige Teileumfänge zur schlanken Versorgung der Automobilmontage: Lehrstuhl für Fördertechnik, Materialfluss, Logistik, Techn. Univ. München.

Daganzo, C. F. (2003). A Theory of Supply Chains. Lecture Notes in Economics and Mathematical Systems: Vol. 526. Berlin, Heidelberg: Springer. Retrieved from http://dx.doi.org/10.1007/978-3-642-18152-8

Daganzo, C. F. (2005). Logistics Systems Analysis (Fourth Edition). Berlin, Heidelberg: Springer-Verlag Berlin Heidelberg. Retrieved from http://site.ebrary.com/lib/alltitles/docDetail.action?docID=10268646

Daganzo, C. F. (2012). A theory of supply chains: Springer Science & Business Media.

Dehdari, P. (2014). Measuring the Impact of Lean Techniques on Performance Indicators in Logistics Operations. Zugl.: Karlsruhe, Karlsruher Institut für Technologie (KIT), Diss., 2013. Wissenschaftliche Berichte des Institutes für Fördertechnik und Logistiksysteme des Karlsruher Instituts für Technologie: Vol. 80. Karlsruhe, Hannover: KIT Scientific Publishing; Technische Informationsbibliothek u. Universitätsbibliothek. Retrieved from http://edok01.tib.uni-hannover.de/edoks/e01fn15/788331981.pdf

Regulation (ec) no 561/2006 of the european parliament and of the council of 15 march 2006 on the harmonisation of certain social legislation relating to road transport and amending council regulations (eec) no 3821/85 and (ec) no 2135/98 and repealing council regulation (eec) no 3820/85., European Parliament and Council 2006.

Furmans, K., & Veit, M. (2013). Models of leveling for lean manufacturing systems. In Handbook of Stochastic Models and Analysis of Manufacturing System Operations (pp. 115–138). Springer.

Gillett, B. E., & Miller, L. R. (1974). A heuristic algorithm for the vehicle-dispatch problem. Operations Research, 22(2), 340–349.

Gini, C. (1921). Measurement of inequality of incomes. The Economic Journal, 31(121), 124–126.

Grassmann, W. K., & Jain, J. L. (1989). Numerical solutions of the waiting time distribution and idle time distribution of the arithmetic GI/G/1 queue. Operations Research, 37(1), 141–150.

Grunewald, M. (2015). Planung von Milkruns in der Beschaffungslogistik der Automobilindustrie: Ein Ansatz zur Integration von Bestandsmanagement und Tourenplanung: Springer-Verlag.

Gudehus, T. (1976). Staueffekte vor Transportknoten. Zeitschrift für Operations Research, 20(6), B207-B252. https://doi.org/10.1007/BF01918395

Güllü, R. (1998). Base stock policies for production/inventory problems with uncertain capacity levels. European Journal of Operational Research, 105(1), 43–51.

Handbook of Stochastic Models and Analysis of Manufacturing System Operations. (2013): Springer.

Harris, F. W. (1913). How many parts to make at once. Factory, the magazine of management, 10(2), 135–136.

Hopp, W. J., & Spearman, M. L. (2004). To pull or not to pull: what is the question? Manufacturing & service operations management, 6(2), 133–148.

Hopp, W. J., & Spearman, M. L. (2011). Factory physics (3. ed., reissued.). Long Grove, Ill.: Waveland Press.

Jacobs, F. R., & Chase, R. B. (2017). Operations and supply chain management (Fifteenth edition). The McGraw-Hill education series. New York, NY: McGraw-Hill Education.

Katsnelson, J., & Kotz, S. (1957). On the upper limits of some measures of variability. Archiv für Meteorologie, Geophysik und Bioklimatologie Serie B, 8(1), 103–107. https://doi.org/10.1007/BF02260299

Kingman, J. F. C., & Atiyah, M. F. (1961). The single server queue in heavy traffic. Mathematical Proceedings of the Cambridge Philosophical Society, 57(04), 902. https://doi.org/10.1017/S0305004100036094

Klug, F. (2010). Logistikmanagement in der Automobilindustrie: Grundlagen der Logistik im Automobilbau: Springer-Verlag.

Liker, J. K. (2004). The Toyota way: 14 management principles from the world's greatest manufacturer. New York, NY: McGraw-Hill. Retrieved from http://www.loc.gov/catdir/bios/mh041/2004300007.html

Liker, J. K., & Meier, D. (2006). The Toyota way fieldbook: A practical guide for implementing Toyota's 4Ps. New York: McGraw-Hill. Retrieved from http://www.loc.gov/catdir/enhancements/fy0634 /2006295987-b.html

Lippolt, C. R., & Furmans, K. (2008). Sizing of heijunka-controlled production systems with unreliable production processes. In Lean business systems and beyond (pp. 11–19). Springer.

Little, J. D. C. (1961). A proof for the queuing formula: L= λ W. Operations Research, 9(3), 383–387.

Mankiw, N. G. (2016). Macroeconomics (Ninth edition). New York: Worth Publishers.

Matzka, J., Di Mascolo, M., & Furmans, K. (2012). Buffer sizing of a heijunka Kanban system. Journal of Intelligent Manufacturing, 23(1), 49–60.

Meyer, A. (2015). Milk Run Design: Definitions, Concepts and Solution Approaches. Karlsruhe, Karlsruher Institut für Technologie (KIT), Diss., 2015.

Montgomery, D. C. (2013). Introduction to statistical quality control (7. ed.). Hoboken, NJ: Wiley.

Ohno, T. (1988). Toyota production system: beyond large-scale production: CRC Press.

Ouyang, Y., & Daganzo, C. (2006). Characterization of the bullwhip effect in linear, time-invariant supply chains: some formulae and tests. Management science, 52(10), 1544–1556.

Queiser, Dr. Hartmut. Anlieferkonzepte in der Automobilindustrie - Ein internationaler Vergleich. In ATZ/MTZ–Konferenz Automobil-Montage. Retrieved from http://www.4flow.de/fileadmin/ user_upload_publikationen/veroeffentlichungen/Anlieferkonzepte_in_der_Automobilindustrie_Dr._Hartmut_Queiser.pdf* (Original work published 2007).

Rother, M., & Shook, J. (2009). Learning to see: Value-stream mapping to create value and eliminate muda (Version 1.4). A lean tool kit method and workbook. Cambridge, Mass.: Lean Enterprise Inst.

Sanz, F. J. G., Semmler, K., & Walther, J. (2007). Die Automobilindustrie auf dem Weg zur globalen Netzwerkkompetenz: Springer.

Schöneberg, T. (2013). Selecting cost-minimal delivery profiles and assessing the impact on cost and delivery schedule stability in area forwarding inbound logistics networks in the automotive industry.

Schöneberg, T., Koberstein, A., & Suhl, L. (2010). An optimization model for automated selection of economic and ecologic delivery profiles in area forwarding based inbound logistics networks. Flexible services and manufacturing journal, 22(3-4), 214–235.

Sergei L. Shmakov. (2011). A Universal Method of Solving Quartic Equations. Volume 71 No. 2 2011, 251-259. International Journal of Pure and Applied Mathematics, 71(2), 251–259.

Silver, E. A., Pyke, D. F., & Peterson, R. (1998). Inventory management and production planning and scheduling: Wiley New York.

Smalley, A., Womack, J., Jones, D., Shook, J., & Ferro, J. (2004). Creating level pull: A lean production-system improvement guide for production-control, operations, and engineering professionals ; a lean toolkit method and workbook (Version 1.0, March 2009). Cambridge, MA: Lean Enterprise Institute.

Tempelmeier, H. (2011). Inventory management in supply networks: problems, models, solutions: Books on Demand Norderstedt.

Varian, H. R. (2014). Intermediate microeconomics: A modern approach (9. ed., internat. student ed.). New York, NY: Norton.

Veit, M. (2010). Modelle und Methoden für die Bestandsauslegung in heijunka-nivellierten Supply Chains: KIT Scientific Publishing.

Wagner, H. M., & Whitin, T. M. (1958). Dynamic version of the economic lot size model. Management science, 5(1), 89–96.

Waite, M. (Ed.). (2012). Paperback Oxford English dictionary (Seventh edition). Oxford: Oxford University Press.

Weber, J. (2012). Logistikkostenrechnung: Kosten-, Leistungs- und Erlösinformationen zur erfolgsorientierten Steuerung der Logistik (3. Auflage). SpringerLink Bücher. Berlin, Heidelberg: Springer Vieweg. Retrieved from http://dx.doi.org/10.1007/978-3-642-25173-3

Wilken, V. (2016). KGS 2017 - Kostensätze Gütertransport Straße (Auflage 2017, neue Ausgabe). Düsseldorf: Verkehrs-Verlag Fischer.

Womack, J. P., Jones, D. T., & Roos, D. (2007). The machine that changed the world: [the story of lean production ; Toyota's secret weapon in the global car wars that is revolutionizing world industry] (1. pb. ed.). Business. New York, NY: Free Press.

Zeidler, E., Grosche, G., Ziegler, V., Ziegler, D., & Bronštejn, I. N. (Eds.). (2013). Springer-Taschenbuch der Mathematik (3., neu bearb. und erw. Aufl. 2013). Wiesbaden: Springer. Retrieved from http://dx.doi.org/10.1007/978-3-8348-2359-5

Appendix – Simulation Input Data

Weight per Unit

	Weight per Unit as a Function of G_{Weight}									
	Heavy High Runners					Heavy Low Runners				
Part No.	0.00	0.25	0.50	0.75	1.00	0.00	0.25	0.50	0.75	1.00
1	61.67	326.33	979.08	2333.76	9250	61.67	35.99	18.69	6.64	0.00
2	61.67	240.55	566.25	1037.17	0.00	61.67	36.10	18.79	6.69	0.00
3	61.67	201.25	411.05	645.39	0.00	61.67	36.20	18.89	6.74	0.00
4	61.67	177.32	327.48	460.94	0.00	61.67	36.31	18.99	6.80	0.00
5	61.67	160.74	274.56	355.03	0.00	61.67	36.42	19.10	6.85	0.00
6	61.67	148.35	237.73	286.83	0.00	61.67	36.53	19.20	6.91	0.00
7	61.67	138.62	210.47	239.49	0.00	61.67	36.64	19.31	6.96	0.00
8	61.67	130.71	189.40	204.85	0.00	61.67	36.75	19.41	7.02	0.00
9	61.67	124.11	172.57	178.48	0.00	61.67	36.87	19.52	7.08	0.00
10	61.67	118.48	158.79	157.78	0.00	61.67	36.98	19.63	7.14	0.00
11	61.67	113.62	147.27	141.13	0.00	61.67	37.10	19.74	7.20	0.00
12	61.67	109.35	137.49	127.47	0.00	61.67	37.22	19.85	7.26	0.00
13	61.67	105.57	129.06	116.08	0.00	61.67	37.33	19.97	7.32	0.00
14	61.67	102.18	121.72	106.44	0.00	61.67	37.45	20.08	7.38	0.00
15	61.67	99.12	115.27	98.18	0.00	61.67	37.58	20.20	7.44	0.00
16	61.67	96.35	109.54	91.04	0.00	61.67	37.70	20.32	7.51	0.00
17	61.67	93.81	104.42	84.81	0.00	61.67	37.82	20.44	7.57	0.00
18	61.67	91.48	99.81	79.32	0.00	61.67	37.95	20.56	7.64	0.00
19	61.67	89.33	95.63	74.46	0.00	61.67	38.07	20.68	7.71	0.00

20	61.67	87.34	91.83	70.12	0.00	61.67	38.20	20.81	7.78	0.00
21	61.67	85.48	88.36	66.23	0.00	61.67	38.33	20.93	7.85	0.00
22	61.67	83.75	85.17	62.72	0.00	61.67	38.46	21.06	7.92	0.00
23	61.67	82.13	82.23	59.54	0.00	61.67	38.59	21.19	7.99	0.00
24	61.67	80.61	79.52	56.65	0.00	61.67	38.72	21.32	8.07	0.00
25	61.67	79.17	76.99	54.01	0.00	61.67	38.86	21.45	8.14	0.00
26	61.67	77.82	74.64	51.59	0.00	61.67	39.00	21.59	8.22	0.00
27	61.67	76.54	72.45	49.36	0.00	61.67	39.13	21.73	8.29	0.00
28	61.67	75.32	70.40	47.30	0.00	61.67	39.27	21.87	8.37	0.00
29	61.67	74.17	68.47	45.40	0.00	61.67	39.42	22.01	8.45	0.00
30	61.67	73.07	66.66	43.63	0.00	61.67	39.56	22.15	8.53	0.00
31	61.67	72.02	64.96	41.99	0.00	61.67	39.70	22.30	8.62	0.00
32	61.67	71.02	63.35	40.46	0.00	61.67	39.85	22.45	8.70	0.00
33	61.67	70.07	61.83	39.03	0.00	61.67	40.00	22.60	8.79	0.00
34	61.67	69.15	60.39	37.69	0.00	61.67	40.15	22.75	8.88	0.00
35	61.67	68.28	59.02	36.43	0.00	61.67	40.30	22.90	8.97	0.00
36	61.67	67.44	57.72	35.25	0.00	61.67	40.45	23.06	9.06	0.00
37	61.67	66.63	56.49	34.14	0.00	61.67	40.61	23.22	9.15	0.00
38	61.67	65.85	55.31	33.09	0.00	61.67	40.77	23.38	9.25	0.00
39	61.67	65.10	54.18	32.10	0.00	61.67	40.93	23.55	9.34	0.00
40	61.67	64.38	53.11	31.16	0.00	61.67	41.09	23.71	9.44	0.00
41	61.67	63.68	52.09	30.28	0.00	61.67	41.25	23.88	9.54	0.00
42	61.67	63.01	51.10	29.43	0.00	61.67	41.42	24.06	9.64	0.00
43	61.67	62.36	50.16	28.64	0.00	61.67	41.59	24.23	9.75	0.00
44	61.67	61.74	49.26	27.88	0.00	61.67	41.76	24.41	9.86	0.00
45	61.67	61.13	48.39	27.15	0.00	61.67	41.93	24.59	9.96	0.00
46	61.67	60.54	47.56	26.46	0.00	61.67	42.11	24.78	10.08	0.00
47	61.67	59.97	46.76	25.80	0.00	61.67	42.28	24.97	10.19	0.00
48	61.67	59.42	45.99	25.18	0.00	61.67	42.46	25.16	10.30	0.00

49	61.67	58.88	45.24	24.58	0.00	61.67	42.65	25.35	10.42	0.00
50	61.67	58.36	44.53	24.00	0.00	61.67	42.83	25.55	10.54	0.00
51	61.67	57.85	43.84	23.45	0.00	61.67	43.02	25.75	10.67	0.00
52	61.67	57.36	43.17	22.93	0.00	61.67	43.21	25.96	10.79	0.00
53	61.67	56.88	42.52	22.42	0.00	61.67	43.40	26.17	10.92	0.00
54	61.67	56.42	41.90	21.94	0.00	61.67	43.60	26.38	11.05	0.00
55	61.67	55.96	41.30	21.47	0.00	61.67	43.80	26.60	11.19	0.00
56	61.67	55.52	40.71	21.02	0.00	61.67	44.00	26.82	11.33	0.00
57	61.67	55.09	40.15	20.59	0.00	61.67	44.21	27.04	11.47	0.00
58	61.67	54.67	39.60	20.18	0.00	61.67	44.41	27.27	11.61	0.00
59	61.67	54.26	39.07	19.78	0.00	61.67	44.63	27.51	11.76	0.00
60	61.67	53.86	38.55	19.39	0.00	61.67	44.84	27.74	11.91	0.00
61	61.67	53.47	38.05	19.02	0.00	61.67	45.06	27.99	12.07	0.00
62	61.67	53.09	37.57	18.66	0.00	61.67	45.28	28.24	12.23	0.00
63	61.67	52.72	37.10	18.32	0.00	61.67	45.51	28.49	12.39	0.00
64	61.67	52.35	36.64	17.98	0.00	61.67	45.74	28.75	12.56	0.00
65	61.67	52.00	36.19	17.66	0.00	61.67	45.97	29.01	12.73	0.00
66	61.67	51.65	35.76	17.35	0.00	61.67	46.21	29.28	12.90	0.00
67	61.67	51.31	35.34	17.04	0.00	61.67	46.45	29.56	13.08	0.00
68	61.67	50.98	34.93	16.75	0.00	61.67	46.69	29.84	13.27	0.00
69	61.67	50.65	34.52	16.47	0.00	61.67	46.94	30.12	13.46	0.00
70	61.67	50.33	34.13	16.19	0.00	61.67	47.20	30.42	13.65	0.00
71	61.67	50.02	33.75	15.93	0.00	61.67	47.46	30.72	13.85	0.00
72	61.67	49.71	33.38	15.67	0.00	61.67	47.72	31.02	14.06	0.00
73	61.67	49.41	33.02	15.42	0.00	61.67	47.99	31.34	14.27	0.00
74	61.67	49.11	32.67	15.17	0.00	61.67	48.26	31.66	14.48	0.00
75	61.67	48.82	32.32	14.94	0.00	61.67	48.54	31.99	14.71	0.00
76	61.67	48.54	31.99	14.71	0.00	61.67	48.82	32.32	14.94	0.00
77	61.67	48.26	31.66	14.48	0.00	61.67	49.11	32.67	15.17	0.00

78	61.67	47.99	31.34	14.27	0.00	61.67	49.41	33.02	15.42	0.00
79	61.67	47.72	31.02	14.06	0.00	61.67	49.71	33.38	15.67	0.00
80	61.67	47.46	30.72	13.85	0.00	61.67	50.02	33.75	15.93	0.00
81	61.67	47.20	30.42	13.65	0.00	61.67	50.33	34.13	16.19	0.00
82	61.67	46.94	30.12	13.46	0.00	61.67	50.65	34.52	16.47	0.00
83	61.67	46.69	29.84	13.27	0.00	61.67	50.98	34.93	16.75	0.00
84	61.67	46.45	29.56	13.08	0.00	61.67	51.31	35.34	17.04	0.00
85	61.67	46.21	29.28	12.90	0.00	61.67	51.65	35.76	17.35	0.00
86	61.67	45.97	29.01	12.73	0.00	61.67	52.00	36.19	17.66	0.00
87	61.67	45.74	28.75	12.56	0.00	61.67	52.35	36.64	17.98	0.00
88	61.67	45.51	28.49	12.39	0.00	61.67	52.72	37.10	18.32	0.00
89	61.67	45.28	28.24	12.23	0.00	61.67	53.09	37.57	18.66	0.00
90	61.67	45.06	27.99	12.07	0.00	61.67	53.47	38.05	19.02	0.00
91	61.67	44.84	27.74	11.91	0.00	61.67	53.86	38.55	19.39	0.00
92	61.67	44.63	27.51	11.76	0.00	61.67	54.26	39.07	19.78	0.00
93	61.67	44.41	27.27	11.61	0.00	61.67	54.67	39.60	20.18	0.00
94	61.67	44.21	27.04	11.47	0.00	61.67	55.09	40.15	20.59	0.00
95	61.67	44.00	26.82	11.33	0.00	61.67	55.52	40.71	21.02	0.00
96	61.67	43.80	26.60	11.19	0.00	61.67	55.96	41.30	21.47	0.00
97	61.67	43.60	26.38	11.05	0.00	61.67	56.42	41.90	21.94	0.00
98	61.67	43.40	26.17	10.92	0.00	61.67	56.88	42.52	22.42	0.00
99	61.67	43.21	25.96	10.79	0.00	61.67	57.36	43.17	22.93	0.00
100	61.67	43.02	25.75	10.67	0.00	61.67	57.85	43.84	23.45	0.00
101	61.67	42.83	25.55	10.54	0.00	61.67	58.36	44.53	24.00	0.00
102	61.67	42.65	25.35	10.42	0.00	61.67	58.88	45.24	24.58	0.00
103	61.67	42.46	25.16	10.30	0.00	61.67	59.42	45.99	25.18	0.00
104	61.67	42.28	24.97	10.19	0.00	61.67	59.97	46.76	25.80	0.00
105	61.67	42.11	24.78	10.08	0.00	61.67	60.54	47.56	26.46	0.00
106	61.67	41.93	24.59	9.96	0.00	61.67	61.13	48.39	27.15	0.00

107	61.67	41.76	24.41	9.86	0.00	61.67	61.74	49.26	27.88	0.00
108	61.67	41.59	24.23	9.75	0.00	61.67	62.36	50.16	28.64	0.00
109	61.67	41.42	24.06	9.64	0.00	61.67	63.01	51.10	29.43	0.00
110	61.67	41.25	23.88	9.54	0.00	61.67	63.68	52.09	30.28	0.00
111	61.67	41.09	23.71	9.44	0.00	61.67	64.38	53.11	31.16	0.00
112	61.67	40.93	23.55	9.34	0.00	61.67	65.10	54.18	32.10	0.00
113	61.67	40.77	23.38	9.25	0.00	61.67	65.85	55.31	33.09	0.00
114	61.67	40.61	23.22	9.15	0.00	61.67	66.63	56.49	34.14	0.00
115	61.67	40.45	23.06	9.06	0.00	61.67	67.44	57.72	35.25	0.00
116	61.67	40.30	22.90	8.97	0.00	61.67	68.28	59.02	36.43	0.00
117	61.67	40.15	22.75	8.88	0.00	61.67	69.15	60.39	37.69	0.00
118	61.67	40.00	22.60	8.79	0.00	61.67	70.07	61.83	39.03	0.00
119	61.67	39.85	22.45	8.70	0.00	61.67	71.02	63.35	40.46	0.00
120	61.67	39.70	22.30	8.62	0.00	61.67	72.02	64.96	41.99	0.00
121	61.67	39.56	22.15	8.53	0.00	61.67	73.07	66.66	43.63	0.00
122	61.67	39.42	22.01	8.45	0.00	61.67	74.17	68.47	45.40	0.00
123	61.67	39.27	21.87	8.37	0.00	61.67	75.32	70.40	47.30	0.00
124	61.67	39.13	21.73	8.29	0.00	61.67	76.54	72.45	49.36	0.00
125	61.67	39.00	21.59	8.22	0.00	61.67	77.82	74.64	51.59	0.00
126	61.67	38.86	21.45	8.14	0.00	61.67	79.17	76.99	54.01	0.00
127	61.67	38.72	21.32	8.07	0.00	61.67	80.61	79.52	56.65	0.00
128	61.67	38.59	21.19	7.99	0.00	61.67	82.13	82.23	59.54	0.00
129	61.67	38.46	21.06	7.92	0.00	61.67	83.75	85.17	62.72	0.00
130	61.67	38.33	20.93	7.85	0.00	61.67	85.48	88.36	66.23	0.00
131	61.67	38.20	20.81	7.78	0.00	61.67	87.34	91.83	70.12	0.00
132	61.67	38.07	20.68	7.71	0.00	61.67	89.33	95.63	74.46	0.00
133	61.67	37.95	20.56	7.64	0.00	61.67	91.48	99.81	79.32	0.00
134	61.67	37.82	20.44	7.57	0.00	61.67	93.81	104.42	84.81	0.00
135	61.67	37.70	20.32	7.51	0.00	61.67	96.35	109.54	91.04	0.00

136	61.67	37.58	20.20	7.44	0.00	61.67	99.12	115.27	98.18	0.00
137	61.67	37.45	20.08	7.38	0.00	61.67	102.18	121.72	106.44	0.00
138	61.67	37.33	19.97	7.32	0.00	61.67	105.57	129.06	116.08	0.00
139	61.67	37.22	19.85	7.26	0.00	61.67	109.35	137.49	127.47	0.00
140	61.67	37.10	19.74	7.20	0.00	61.67	113.62	147.27	141.13	0.00
141	61.67	36.98	19.63	7.14	0.00	61.67	118.48	158.79	157.78	0.00
142	61.67	36.87	19.52	7.08	0.00	61.67	124.11	172.57	178.48	0.00
143	61.67	36.75	19.41	7.02	0.00	61.67	130.71	189.40	204.85	0.00
144	61.67	36.64	19.31	6.96	0.00	61.67	138.62	210.47	239.49	0.00
145	61.67	36.53	19.20	6.91	0.00	61.67	148.35	237.73	286.83	0.00
146	61.67	36.42	19.10	6.85	0.00	61.67	160.74	274.56	355.03	0.00
147	61.67	36.31	18.99	6.80	0.00	61.67	177.32	327.48	460.94	0.00
148	61.67	36.20	18.89	6.74	0.00	61.67	201.25	411.05	645.39	0.00
149	61.67	36.10	18.79	6.69	0.00	61.67	240.55	566.25	1037.17	0.00
150	61.67	35.99	18.69	6.64	0.00	61.67	326.33	979.08	2333.76	9250

Mean Demand per Unit

	Mean Demand as a Function of $G_{Quantity}$				
Part No.	0.00	0.25	0.50	0.75	1.00
1	0.33	1.76	5.29	12.61	50.00
2	0.33	1.30	3.06	5.61	0.00
3	0.33	1.09	2.22	3.49	0.00
4	0.33	0.96	1.77	2.49	0.00
5	0.33	0.87	1.48	1.92	0.00
6	0.33	0.80	1.29	1.55	0.00
7	0.33	0.75	1.14	1.29	0.00
8	0.33	0.71	1.02	1.11	0.00
9	0.33	0.67	0.93	0.96	0.00
10	0.33	0.64	0.86	0.85	0.00
11	0.33	0.61	0.80	0.76	0.00
12	0.33	0.59	0.74	0.69	0.00
13	0.33	0.57	0.70	0.63	0.00
14	0.33	0.55	0.66	0.58	0.00
15	0.33	0.54	0.62	0.53	0.00
16	0.33	0.52	0.59	0.49	0.00
17	0.33	0.51	0.56	0.46	0.00
18	0.33	0.49	0.54	0.43	0.00
19	0.33	0.48	0.52	0.40	0.00
20	0.33	0.47	0.50	0.38	0.00
21	0.33	0.46	0.48	0.36	0.00
22	0.33	0.45	0.46	0.34	0.00
23	0.33	0.44	0.44	0.32	0.00
24	0.33	0.44	0.43	0.31	0.00

25	0.33	0.43	0.42	0.29	0.00
26	0.33	0.42	0.40	0.28	0.00
27	0.33	0.41	0.39	0.27	0.00
28	0.33	0.41	0.38	0.26	0.00
29	0.33	0.40	0.37	0.25	0.00
30	0.33	0.39	0.36	0.24	0.00
31	0.33	0.39	0.35	0.23	0.00
32	0.33	0.38	0.34	0.22	0.00
33	0.33	0.38	0.33	0.21	0.00
34	0.33	0.37	0.33	0.20	0.00
35	0.33	0.37	0.32	0.20	0.00
36	0.33	0.36	0.31	0.19	0.00
37	0.33	0.36	0.31	0.18	0.00
38	0.33	0.36	0.30	0.18	0.00
39	0.33	0.35	0.29	0.17	0.00
40	0.33	0.35	0.29	0.17	0.00
41	0.33	0.34	0.28	0.16	0.00
42	0.33	0.34	0.28	0.16	0.00
43	0.33	0.34	0.27	0.15	0.00
44	0.33	0.33	0.27	0.15	0.00
45	0.33	0.33	0.26	0.15	0.00
46	0.33	0.33	0.26	0.14	0.00
47	0.33	0.32	0.25	0.14	0.00
48	0.33	0.32	0.25	0.14	0.00
49	0.33	0.32	0.24	0.13	0.00
50	0.33	0.32	0.24	0.13	0.00
51	0.33	0.31	0.24	0.13	0.00
52	0.33	0.31	0.23	0.12	0.00
53	0.33	0.31	0.23	0.12	0.00

54	0.33	0.30	0.23	0.12	0.00
55	0.33	0.30	0.22	0.12	0.00
56	0.33	0.30	0.22	0.11	0.00
57	0.33	0.30	0.22	0.11	0.00
58	0.33	0.30	0.21	0.11	0.00
59	0.33	0.29	0.21	0.11	0.00
60	0.33	0.29	0.21	0.10	0.00
61	0.33	0.29	0.21	0.10	0.00
62	0.33	0.29	0.20	0.10	0.00
63	0.33	0.28	0.20	0.10	0.00
64	0.33	0.28	0.20	0.10	0.00
65	0.33	0.28	0.20	0.10	0.00
66	0.33	0.28	0.19	0.09	0.00
67	0.33	0.28	0.19	0.09	0.00
68	0.33	0.28	0.19	0.09	0.00
69	0.33	0.27	0.19	0.09	0.00
70	0.33	0.27	0.18	0.09	0.00
71	0.33	0.27	0.18	0.09	0.00
72	0.33	0.27	0.18	0.08	0.00
73	0.33	0.27	0.18	0.08	0.00
74	0.33	0.27	0.18	0.08	0.00
75	0.33	0.26	0.17	0.08	0.00
76	0.33	0.26	0.17	0.08	0.00
77	0.33	0.26	0.17	0.08	0.00
78	0.33	0.26	0.17	0.08	0.00
79	0.33	0.26	0.17	0.08	0.00
80	0.33	0.26	0.17	0.07	0.00
81	0.33	0.26	0.16	0.07	0.00
82	0.33	0.25	0.16	0.07	0.00

83	0.33	0.25	0.16	0.07	0.00
84	0.33	0.25	0.16	0.07	0.00
85	0.33	0.25	0.16	0.07	0.00
86	0.33	0.25	0.16	0.07	0.00
87	0.33	0.25	0.16	0.07	0.00
88	0.33	0.25	0.15	0.07	0.00
89	0.33	0.24	0.15	0.07	0.00
90	0.33	0.24	0.15	0.07	0.00
91	0.33	0.24	0.15	0.06	0.00
92	0.33	0.24	0.15	0.06	0.00
93	0.33	0.24	0.15	0.06	0.00
94	0.33	0.24	0.15	0.06	0.00
95	0.33	0.24	0.14	0.06	0.00
96	0.33	0.24	0.14	0.06	0.00
97	0.33	0.24	0.14	0.06	0.00
98	0.33	0.23	0.14	0.06	0.00
99	0.33	0.23	0.14	0.06	0.00
100	0.33	0.23	0.14	0.06	0.00
101	0.33	0.23	0.14	0.06	0.00
102	0.33	0.23	0.14	0.06	0.00
103	0.33	0.23	0.14	0.06	0.00
104	0.33	0.23	0.13	0.06	0.00
105	0.33	0.23	0.13	0.05	0.00
106	0.33	0.23	0.13	0.05	0.00
107	0.33	0.23	0.13	0.05	0.00
108	0.33	0.22	0.13	0.05	0.00
109	0.33	0.22	0.13	0.05	0.00
110	0.33	0.22	0.13	0.05	0.00
111	0.33	0.22	0.13	0.05	0.00

112	0.33	0.22	0.13	0.05	0.00
113	0.33	0.22	0.13	0.05	0.00
114	0.33	0.22	0.13	0.05	0.00
115	0.33	0.22	0.12	0.05	0.00
116	0.33	0.22	0.12	0.05	0.00
117	0.33	0.22	0.12	0.05	0.00
118	0.33	0.22	0.12	0.05	0.00
119	0.33	0.22	0.12	0.05	0.00
120	0.33	0.21	0.12	0.05	0.00
121	0.33	0.21	0.12	0.05	0.00
122	0.33	0.21	0.12	0.05	0.00
123	0.33	0.21	0.12	0.05	0.00
124	0.33	0.21	0.12	0.04	0.00
125	0.33	0.21	0.12	0.04	0.00
126	0.33	0.21	0.12	0.04	0.00
127	0.33	0.21	0.12	0.04	0.00
128	0.33	0.21	0.11	0.04	0.00
129	0.33	0.21	0.11	0.04	0.00
130	0.33	0.21	0.11	0.04	0.00
131	0.33	0.21	0.11	0.04	0.00
132	0.33	0.21	0.11	0.04	0.00
133	0.33	0.21	0.11	0.04	0.00
134	0.33	0.20	0.11	0.04	0.00
135	0.33	0.20	0.11	0.04	0.00
136	0.33	0.20	0.11	0.04	0.00
137	0.33	0.20	0.11	0.04	0.00
138	0.33	0.20	0.11	0.04	0.00
139	0.33	0.20	0.11	0.04	0.00
140	0.33	0.20	0.11	0.04	0.00

141	0.33	0.20	0.11	0.04	0.00
142	0.33	0.20	0.11	0.04	0.00
143	0.33	0.20	0.10	0.04	0.00
144	0.33	0.20	0.10	0.04	0.00
145	0.33	0.20	0.10	0.04	0.00
146	0.33	0.20	0.10	0.04	0.00
147	0.33	0.20	0.10	0.04	0.00
148	0.33	0.20	0.10	0.04	0.00
149	0.33	0.20	0.10	0.04	0.00
150	0.33	0.19	0.10	0.04	0.00